AF226130

Beyond the Quantum Fringe

2026 REVISION

A New Physics of the Sub-Quantum Fabric and the Technology it makes in a Post-Quantum Age

"Science is about exploring the possibility of anything, to discover and learn about the probability of everything."

Written and illustrated by

Chris Daniels

Published by DCscienceworks - Fringe Science Publishers 2026

Contents

AN EXPLANATION OF FRINGE SCIENCES

"Science is about exploring the possibility of anything, to discover and learn about the probability of everything."

Fringe sciences are fields of study that stretch beyond the limits of mainstream theories and accepted methods. Some eventually become part of established science, while others remain on the edge because the ideas are ahead of their time, the tools to test them are still developing, or the wider scientific community has not yet caught up. Unlike pseudoscience, fringe science is rooted in genuine inquiry and known principles, but it is not afraid to push further. Its purpose is to explore ideas that challenge or expand the boundaries of what we know and understand.

Many subjects once dismissed as fringe have since become central to modern science. These include advanced theories of mind and brain function, the study of the superconscious, the search for the Higgs boson (popularly labelled the "God Particle"), cybernetics, synthetic A.I. with the potential for self-awareness, black holes, teleportation concepts, time dilation and time travel theories, quantum physics, and even relativity. Relativity, despite being one of the pillars of modern physics today, was originally viewed as radical and still generates debate in some scientific circles.

What unites people working in these fields is a shared drive to expand our understanding of the world, the universe, and the eternal Omniverse. They seek a broader picture that explains how everything fits together, what lies beyond the limits of our instruments and assumptions, and how we fit within a much greater reality. The aim is not to create controversy, but to encourage discovery and genuine advancement.

QUANTUM IS AN UNKNOWN MEASURE

The word *quantum* is used frequently today. It appears in everything from "quantum leaping" to "quantum healing" and "quantum consciousness," often to suggest something impossibly small or impossibly vast. The original meaning, however, is far more precise.

Quantum comes from the Latin *quantus*, meaning "how much" or "how great." It's a question, not an answer. It refers to a quantity that has not yet been measured. The related phrase *quantus mensio* translates roughly to "what is the measure, the quantity?"

In this sense, quantum does not mean small or large. It simply means the measure is unknown. Quantum physics seeks to uncover those unknowns, while quantum dynamics continues to challenge, puzzle and reshape how we think about physical reality.

Even in mainstream physics, the term reflects uncertainty. It describes the limits of what we can measure or predict. Seen this way, the quantum domain represents the true frontier of science, a reminder that the deepest mysteries often sit just beyond the threshold of our current understanding.

This book is for anyone curious enough to explore the edge where science and philosophy meet. It's a space where answers are incomplete, where the questions become more interesting, and where imagination matters as much as measurement. Whether your interest lies in how the universe and Omniverse might truly function, or whether your perspective is more theological, pantheist or deist, there is likely something here worth thinking about, and perhaps something that will speak to you.

Exploring the fringes of science can be both thrilling and unsettling. From theoretical possibilities such as time travel and clean, perpetual mass–energy, to the deeper building blocks and dark structures of the Omniverse that hint at multiple dimensions and alternate realities, this book moves deliberately into territory designed to stretch the mind. The intention is not to overwhelm, but to offer ideas that may support future progress and open the door to new questions, discussions and solutions.

The material draws from more than forty years of my work, including articles, studies, observations, theses, dissertations and propositions. It blends theoretical and applied science with philosophy in an effort to step beyond what is commonly accepted and into new territory, so others may continue the journey from a higher vantage point.

I am not claiming any of this as absolute fact. These are concepts, observations and possibilities. But history has shown that even the most unconventional ideas can spark real advancements when they are explored with honesty and rigour.
Fringe inquiry is not the enemy of science. It's often the spark that drives it forward.

Science should not be an exclusive club. You should not need a specialised degree to take part in the conversation, especially when the biggest questions, existence, consciousness, purpose and the future of humanity, belong to all of us.

While some sections reach into deeper theory, the language has been kept clear and approachable so that any curious reader can follow the flow.

My hope is that open, constructive dialogue, free from hierarchy and bias, can lead us toward knowledge that improves the human condition and life on Earth, and broadens our horizons far beyond it. If we are willing to explore together, scientifically, philosophically, emotionally and imaginatively, we may discover better ways to evolve as a species and contribute more meaningfully to the greater fabric of reality we are part of.

A CAUTIONARY TALE

To understand the heart of this journey, you will first need to walk through a series of chapters that lay the groundwork. Some sections may feel slow or slightly repetitive, but each carries a necessary piece of the puzzle. These early steps build a foundation for insights that only fully make sense once all the elements are in place.

If you stay with it, you may begin to notice connections forming in unexpected ways. What starts as context gradually reveals a broader pattern. And when that pattern comes into view, the journey often proves not only worthwhile, but far more intriguing than it first appears.

A FOUNDING PHILOSOPHY

Explore, discover and learn, to ultimately create

People often ask how I know what I know, or where I learned it.

The answer is simple. Anyone can learn these things. Knowledge is everywhere if you choose to look and listen. It lives in books and lectures, in conversations with elders, in the way people move through a busy street, and in the quiet persistence of ants building a home in the desert. It exists in classrooms, libraries, jungles, oceans and star-filled skies. Learning never truly stops. And the purpose of learning is to gather the understanding required to create.

Creation takes many forms, far more than we usually recognise. For me, meaningful creation grows from exploration and discovery. It's what happens when knowledge and wisdom converge into something new. I've often said that if you learn enough to fill a thousand galaxies, only then might you understand how to create a perfect acorn. And consider what a single acorn becomes.

What sets humanity apart is our capacity for complex creative imagination. Animals imagine in simple ways. A dog may see a steak and imagine eating it. But we imagine beyond instinct. We contemplate biology, question the cosmos, and consider outcomes long before events unfold. We visualise multiple paths and choose the one we believe leads to the greatest possibility.

Through imagination, we create. At our best, that creative power can feel almost divine. But, as Stan Lee reminded us through Spider-Man's Uncle Ben:

"With great power comes great responsibility."

This philosophy sits at the heart of everything in this book. Explore widely. Discover boldly. Learn relentlessly. Then create something worthy of the journey.

THE AMBITIOUS MISSIONARY

My intention here is ambitious. I aim to encourage new ways of thinking, spark imagination, and open the door to deeper exploration at the intersection of science and philosophy. If humanity is to continue advancing, these two domains can no longer operate as separate territories. They must inform and enrich one another.

Too often, human knowledge is divided into camps. On one side sits orthodox academia, cautious and structured. On the other stands the independent thinker, the intuitive theorist, the explorer who recognises patterns and possibilities long before they are formally acknowledged. History is filled with breakthroughs that began on the fringes. Many ideas considered unorthodox at first later became accepted science once the tools existed to test or measure them.

This book looks beyond the orthodox and into areas often dismissed: paraphysics, metaphysics and pataphysics, the philosophy of imagined possibilities. When approached with honesty and discipline, these fields can reveal insights that mainstream methods may overlook. Sometimes the unconventional path simply reaches the horizon earlier.

Discernment, of course, is essential. Pseudoscience, driven by manipulation or self-serving claims, has nothing to do with genuine inquiry. But closing the door on all unconventional ideas is equally limiting.

If we refuse to explore the edges, we risk missing discoveries that could transform our understanding of reality.

Imagine demonstrating that thought behaves as a physical property, with its own form of velocity and momentum.
Imagine uncovering deeper connections between time, gravity and consciousness that current models only hint at.

Ideas once confined to science fiction, such as time travel, alternate realities or collective consciousness, are now being re-examined through quantum computing, brain imaging, advanced A.I. and emergent theories in physics.

For decades, I've proposed that thought may possess velocity, potentially making the movement of light seem slow by comparison. I've also suggested that the Gem-Ion, the geometric eternally metamorphic ion, may be a fundamental building block of the universe and Omniverse. These are bold ideas, but they sit within a growing global effort to push physics beyond classical boundaries and explore deeper layers of structure in reality.

If we remain open to new ideas, while grounding them in logic, observation and rigorous thinking, we give ourselves the chance to recognise these dynamics not as wild speculation, but as potential physical systems that future science may measure, influence and apply. This is the heart of real exploration: to question, to imagine, to discover, and to create.

As the old saying goes:
"There is more to life than meets the eye."

CONSTRUCT OF THE UNIVERSE

Abandon the rules and find the Answers

The idea of bringing together universal building blocks, gravitational behaviour, energetic and electromagnetic dynamics, bending space, faster-than-light travel, time travel, black holes, wormholes and parallel universes into one coherent *Unification Theory of Everything* has occupied thinkers for decades. There have been exciting breakthroughs and promising theories, yet the final answer always seems to slip just beyond our reach. Sooner or later, we hit a conceptual wall that stalls momentum and keeps the deeper truth out of sight.

We know everything's connected. We're connected to our universe and to the countless universes of the eternal Omniverse. But the idea that everything can be reduced to one neat singularity doesn't ring true. What feels more realistic is a collaborative, interwoven system, not a single point, but a network of relationships.

Since the early 1970s, and more formally since 1989, I've been working on an alternative construct. It's an idea that's only now beginning to surface quietly within wider scientific conversation. For decades, science has circled around this concept without fully stepping into it. This construct may not only help break through the wall that has stalled progress, but also point towards a new and encouraging direction.

The core problem is that we keep searching for answers inside the same box of rules that limits our thinking. When we insist on working only within today's definitions and assumptions, we restrict our own vision. We keep trying to solve an incomplete puzzle using pieces drawn from the same cramped toolkit.

What I'm inviting you to do is step outside that confinement. Allow yourself a little freedom. Imagine what becomes possible when we move beyond today's parameters and rigid laws of physics.

Ideas that seem outlandish right now may, in hindsight, turn out to be simple, rational, and even obvious.

Some readers may be tempted to dismiss these ideas as drifting into metaphysics or "tinfoil hat" territory. If that reaction arises, I'd encourage you to hold judgement for a moment. Set aside bias, professional pride, or reflexive scepticism. This journey may reveal something genuinely valuable, including new ways to unify the universe and the multi-dimensional Omniverse. Along the way, we may also stumble upon practical insights, such as advanced gravitational management or more efficient ways to generate and apply energy.

In genuine exploration, there's no room for ego. If the aim is real progress for humanity, then we need to loosen our grip on what we think we already know and approach these questions as though we're starting again. Real learning often begins with recognising how little we truly understand.

One of the few things we may have grasped so far, drawing from Newton, Maxwell, Tesla, Einstein and Faraday, is that time isn't merely a measure. It's a dimensional state of existence. Yet our physical world isn't actually divided into three separate spatial dimensions plus time.

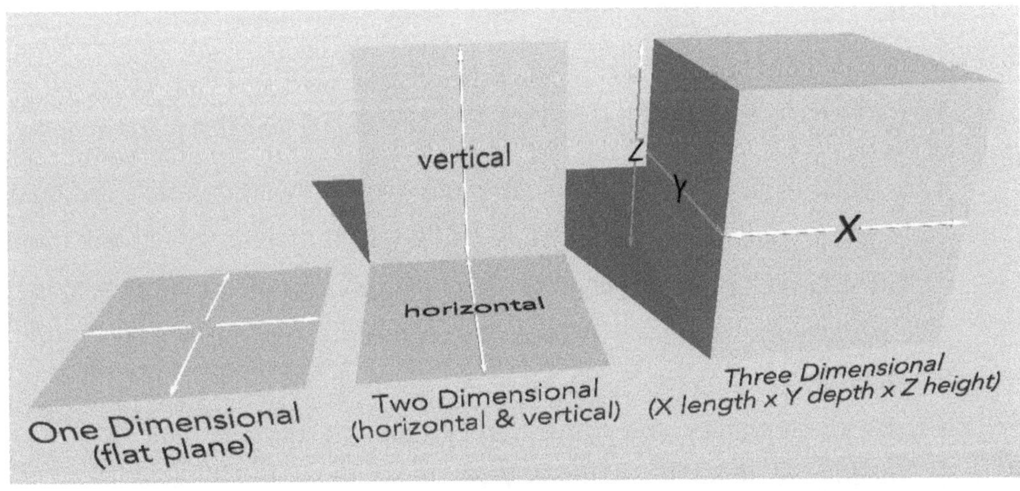

What we call one-dimensional (flat plane), two-dimensional (horizontal and vertical), and three-dimensional (X length, Y depth and Z height) are simply different measures of distance within one unified dimensional realm: the realm of omni-directional distance.

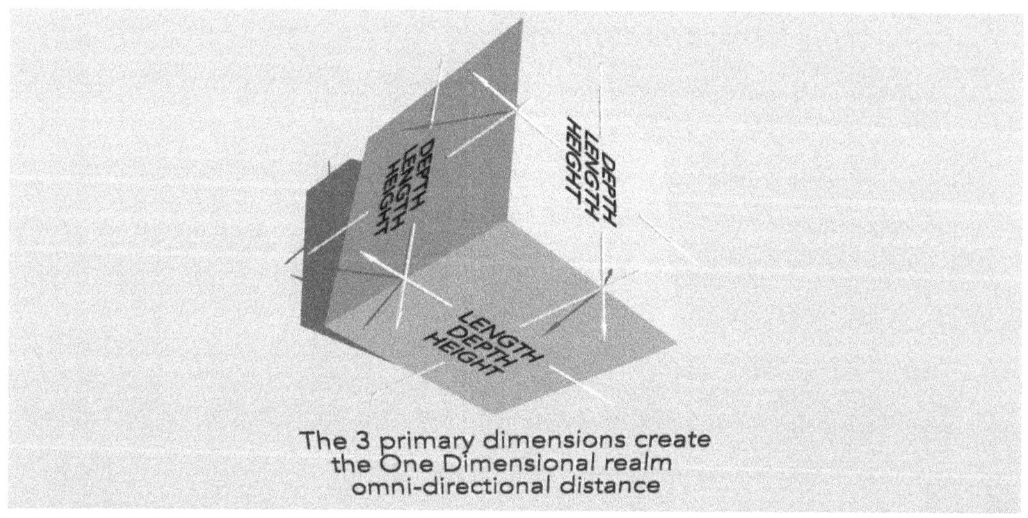

The 3 primary dimensions create
the One Dimensional realm
omni-directional distance

X, Y and Z aren't distinct worlds. They're simply ways of measuring distance. This becomes clearer as we move forward. Ironically, by insisting on the idea of "three dimensions," we've boxed ourselves into definitions that never truly reflect reality. Attempts to stretch this into *multiple dimensions* have often made things more confusing than illuminating. What is simple has been rendered unnecessarily complex. So let's clarify what we mean when we talk about a dimension.

DIMENSION: DEFINITIONS AND INTERPRETATIONS

The word *dimension* is used in many ways. Mathematics, geometry, architecture and everyday language all apply it differently. In the context of the Omniverse, however, four core interpretations are most relevant:

1. A measure of spatial extent or magnitude, such as width, height or length.
2. In mathematics, the minimum number of independent coordinates needed to specify a point in space.

3. In physics, a fundamental measurable property, such as mass, distance or time.
4. A realm of existence, as in alternate dimensions or states of being.

Physicists have long labelled time as a "fourth dimension", yet time behaves nothing like height, length or depth. Time is better understood as an omni-dimensional realm of existence rather than a spatial axis.

For our purposes, the focus is primarily on the fourth definition.

Within that scope, we can recognise six primary dimensional realms: Time, Space, Distance, Thought, Eternity (or Perpetuity), and Existence.

These realms aren't always quantifiable, but they're unquestionably real. We don't need experiments to prove their existence; they're woven directly into perception and experience. By contrast, other dimensions that people speculate about may well exist, but we currently lack the tools to measure or interact with them.

That doesn't make them impossible. It simply places them beyond today's reach. Exploring these unknowns may one day help us understand not only our current theories, but entirely new concepts we haven't yet imagined.

REDUCING THEORY TO PRACTICE

The discussions ahead are theoretical propositions, but I'll keep them grounded in what has plausibility and potential for real-world application.

Because matter plays a central role, you may find it useful later to refer to the subsection *"Construction and Structure of Gem-Ions"* in the chapter *"The Theory of Gem-Ions."* That section explores the building blocks of matter in greater detail. For now, we'll begin with the simplicity of the physical world.

And yes, I say *simplicity* deliberately. While the universe appears overwhelmingly complex, most things can be traced back to simple underlying principles. The construct of matter, the universe and the Omniverse is no exception.

To some, these ideas may sound like fantasy, straying too far from what's considered acceptable science. But it's worth remembering what quantum physics once did to our understanding of reality. Boundaries move. What once seemed impossible can become tomorrow's accepted truth.

As the peculiar saying goes:
"Science is what we do when we don't know what we're doing."

So yes, philosophy has a place here. Perhaps even an essential one.

So, as they say in the classics:
"Strap in and enjoy the ride!"

THE PUZZLE OF ILLUSIONARY REALITY

Works by M. C. Escher, 1898–1972, Master of Illusionary Images

There are three fundamental illusions of our reality, as I see them. They're not illusions in a mystical sense, but the filters, limits and interpretations that our biology brings to the universe itself.

One: our perception of matter.
Two: our perception of light and dark, including the visible and invisible spectrum.
Three: our perception of time and space.

These aren't abstract philosophical riddles. They're the direct result of how our nervous system, sensory organs and cognitive architecture construct the world we think we inhabit. What we call reality isn't a direct experience of the universe. It's a *constructed interpretation*, assembled by the brain from incomplete and selective information.

A fish, an eagle and a bat all inhabit the same planet, yet experience entirely different realities. A fish interprets vibration, pressure and refracted light through water. An eagle sees colour in ultraviolet ranges we can't detect. A bat navigates by acoustic reflection rather than sight. Their realities differ because their sensory systems differ.
Humans are no exception.

Our perception is stitched together from the signals our senses collect and the interpretations our brain applies. This includes sight, sound, touch, taste and smell, along with emotional, memory-driven and psychological influences. Modern neuroscience has shown that the brain doesn't simply record the world. It predicts it. It fills gaps, adjusts context, filters information and sometimes even invents details the senses never received.

This means two people standing side by side, looking at the same object, aren't having the same experience. They're constructing two different internal realities. Those differences become even more pronounced across culture, upbringing, emotional state, memory and learned bias.

Imagine fifty, a hundred, or even a thousand people arranged in a perfect circle around a giant multi-spectral geodesic sphere. Each would see it differently. Their position, line of sight, retinal sensitivity, the way their brain processes contrast and movement, and even their current mood would alter the experience. If the sphere began to rotate, the variations would multiply again. Even when two people look at the same face of the sphere at the same moment, they're still perceiving it through different sensory systems and different mental frameworks.

I refer to this visual and perceptual phenomenon as;
The Illusive Ocularium.

When viewing the Ocularium, every person is experiencing a different sphere, because their biology and psychology compel them to. Their eyes aren't the same. Their neural networks aren't the same. Their memories and emotional states aren't the same.

The Illusive Ocularium

The only way two individuals could experience the exact same perception would be if they shared the same sensory organs, the same neural pathways and the same subjective awareness. Not impossible, but extremely unlikely.

Throughout this book, you'll encounter theories and propositions that explore the gap between the universe itself and the version of it our minds allow us to perceive. Understanding these perceptual filters isn't just philosophy. It's essential to understanding physics, consciousness and the structure of existence itself. Because if we can identify the illusions, we may also begin to glimpse what lies beyond them.

Our perception of matter is shaped by a multi-layered process that works so seamlessly we mistake the final composite for reality itself. Yet what we perceive isn't matter in its true form. It's the integrated illusion produced when several independent layers converge in a single act of observation.

A useful analogy comes from the 3D picture books and comics popular in the 1950 & 60s. The page contained three offset images, one printed in black and the others in red and blue or green. Viewed without glasses, the image appeared distorted and confusing. But once you put on the 3D glasses, each eye filtered a specific colour. The brain then merged the images into a single scene that appeared coherent and three-dimensional.

The illusion felt real. Yet the unfiltered image was just as real. So which version was the truth? In a sense, both are.

Matter behaves in much the same way.
What we experience as solid objects are tangible illusions arising from

multiple layers of physical reality synchronising at the moment of observation. At minimum, four layers, including time, must align to produce an object that appears fully formed and stable.

Now imagine those layers separated. What would each look like? Where would it reside? Which realm would it belong to? And what would matter become if we observed each layer independently rather than all at once?

These aren't idle philosophical questions. They sit at the foundation of how matter behaves, how energy interacts with it, and how our senses and neurological processes shape what we accept as reality.

PRIMARY LAYERS OF MATTER

Material, Prothereal & Ethereal

As I understand it, matter expresses itself across three primary registrations or layers:

Material
Solid matter, including liquids and gases. This is the realm we physically touch and measure.

Prothereal
The subtle framework of space, the ghost-structure of distance, the lattice of time, and the field in which physical processes unfold. It isn't empty. It's an active, dynamic medium.

Ethereal
Non-conforming matter. Sometimes loosely labelled anti-matter, but better understood as matter operating beyond our sensory range and current scientific tools.

Much of what we assume isn't matter is simply matter occupying one of these other layers. Some refer to them as planes, dimensions or alternate realities. While not entirely wrong, those terms lack the precision needed to describe how these layers overlap and interact.

If we loosen our attachment to rules we currently treat as unbreakable, it becomes almost obvious that each layer contributes to the greater whole. To grow scientifically and philosophically, we need a deeper understanding of how these layers interweave into the fabric of the universe and the wider Omniverse.

Layers within Ourselves

We aren't separate from these layers. We're built from them.

- Our physical body belongs primarily to the **Material** layer.
- Our subconscious mind, emotions and psychological architecture operate largely within the **Prothereal** layer. This is where much of our intuition, memory processing and perceptual bias forms. Neurologically, this is where billions of signals are filtered before reaching conscious awareness.
- The deepest part of us belongs to the **Ethereal** layer. This isn't mysticism. It's an acknowledgment that creativity, originality, sudden insight and the emergence of ideas from nowhere may arise from structures of reality our biology isn't designed to perceive directly.

Together, these layers form our complete self.

A Dimensional State of Existence

The Material, Prothereal and Ethereal layers form a unified construct of existence, supported by the major realms of Time, Space, Distance, Thought and Eternity (Perpetuity).

To truly understand the universe, we must recognise how these realms interact and consider what becomes possible when they're manipulated, bypassed or reconfigured.

In everyday life, even walking from the kitchen to the bedroom depends on time, space and distance, yet we rarely think about it. If any one of those realms dissolved, the nature of movement would change entirely.

If distance no longer mattered, you could appear anywhere in time. If space no longer mattered, you could exist in multiple locations simultaneously.
If time and space both collapsed, distance would vanish, and arrival would be instantaneous. Whether it's the next room or another galaxy, the effect would be the same.

This echoes Einstein's description of curved space. Space and time bend. If they can bend, they can fold. And if they can fold, even vast distances can become points of contact.

Physics already hints at this. In quantum dynamics, particles exhibit superposition and non-locality. A particle can exist in multiple states at once. It can occupy multiple positions until measured. It can influence another particle across vast distances with no detectable transmission path, through what we call entanglement.
These behaviours show that distance, location and even time are far more fluid than our senses suggest. The universe already behaves in ways that support dimensional flexibility. We're only just beginning to understand what that means.

OUR ILLUSIONARY PERCEPTION OF LIGHT & DARK

Light and darkness feel like simple opposites. One reveals, the other obscures. But in reality, the picture is far more complex.

What we call darkness may not be the absence of light at all, but the limit of our sensory biology, our instruments and our long-conditioned assumptions.

When we look out into the night sky, we see a black backdrop pierced by stars. It's natural to conclude that space is mostly empty and mostly dark. But that perception isn't a true reading of the universe. It's a narrow, filtered interpretation created by the tiny fraction of the electromagnetic spectrum our eyes can detect.

Science has tried to name the *missing* parts of the cosmos with labels like dark matter, anti-matter and dark energy. These terms aren't explanations. They're placeholders for what we don't yet understand. Dark matter, for instance, isn't dark because it's sinister. It's dark because we can't see it. It doesn't interact with light in ways our instruments can currently detect. Anti-matter isn't "anti" in a moral sense either. It simply behaves differently when interacting with ordinary matter.

Even these labels can mislead us, because they frame the universe around what light reveals, rather than what our perception hides.

Light We Cannot See

The electromagnetic spectrum is vast, stretching from extremely low-frequency radio waves to high-energy gamma radiation. Human eyes can detect only a tiny slice of that spectrum, roughly between 400 and 700 nanometres, sitting between infra-red and ultraviolet.

That means most light in the universe is invisible to us.

With radio telescopes, microwave detectors, X-ray observatories and infra-red imaging, astronomers have already revealed structures completely hidden in visible light:

- Dust-obscured galaxies.
- Nebulae glowing in infra-red.
- Jets of plasma shooting from quasars.
- Filaments of cosmic microwave radiation.
- Entire clusters visible only in X-ray wavelengths.

If, overnight, our eyes evolved to see even twice our current visible range, the night sky would become unrecognisable. Much of what we now call darkness would vanish. Space wouldn't look empty at all. It would likely appear saturated with colours no human has ever seen.

We're essentially standing in a cosmic art gallery with the lights off, describing the paintings from the occasional glint caught by a torch beam.

Sound as a Parallel to Light

The same principle is obvious with sound. Humans hear only a narrow mid-band of the acoustic spectrum, while whales, bats, elephants and even insects communicate far above or below our audible range. The sounds exist whether or not we can hear them.

Reality doesn't shrink to fit our senses. Our senses shrink the reality we're able to notice.

Are We Surrounded by Invisible Life or Processes?

This raises an intriguing question.
If most of the universe is invisible to us, what else might exist beyond our perceptual limits?

We already know that:
- Life exists at molecular and quantum scales,
- energies constantly pass through us that we can't see or feel,
- and, forces like gravity, magnetism and quantum field fluctuations shape everything we experience, yet remain deeply mysterious.

If our senses expanded, or if our technology matured far enough, we might discover forms of structure, behaviour or even life-like systems operating in ranges we've never explored.

This isn't mysticism. It's a straightforward recognition of biological limitation. Every species has a tuned perceptual band. Bees see ultraviolet. Snakes detect heat. Bats navigate with ultrasound. Every creature, including us, lives inside its own sensory bubble.
Why would humans be the one exception with perfect perception?
We're not. We're just clever at pretending we are.

The Illusion of Cosmic Darkness

So when we measure cosmic distances, calculate expansion models or infer universal origins, we're doing it with a restricted data set. It's like trying to describe a mountain range while wearing sunglasses, ear muffs and looking through a keyhole. We can still make progress, but only within the limits of our tools.

As our perceptual reach improves, through better detectors, telescopes and technologies, our picture of the universe shifts. It always has, and it always will.

The dark we see isn't an enemy or a void. Much of it is simply light in another wavelength, or matter in another state, sitting just beyond the borders of our sensory capacity.
Darkness is often nothing more than undiscovered light.

OUR ILLUSIONARY PERCEPTION OF TIME & SPACE

Time is often treated as something that exists on its own, as if it were a separate force flowing past us. In practice, what we call time is largely a way of measuring movement through space.

When we ask, "How long did that take?" and answer, "about a day," we're really describing how far something has travelled across a span of distance.

A day is the time it takes a point on Earth to complete one rotation relative to the sun. A year makes this even clearer. It isn't an abstract interval. It's the distance Earth travels along its orbital path, from one position back to the same position again.

Seen this way, time isn't a separate thing. It's the way we record how objects, including ourselves, move along, across, around or through space.

If the dimensional realm we call time didn't exist, our relationship with space would change completely. In principle, we could move between any two points instantly, regardless of distance. In such a state, ideas like before and after would lose their meaning.

Most people imagine time as a straight line, a one-way road with the past behind us, the future ahead of us and a moving present carrying us along. It's a useful picture, but not a complete one.

A more accurate way to imagine it is like being immersed in an infinite, omni-directional ocean. Every drop in that ocean represents a moment or event. All of those drops exist together within a single continuum.
From our limited viewpoint, we experience that ocean as a current flowing in one direction. But beneath that perception, the ocean itself has no preferred flow. Every moment still exists. We just haven't learned how to navigate between them freely.

In the chapter *A Concept of Non-Linear Time Travel*, I expand on this idea using the image of a vast sphere of time made up of trillions of tiny bubbles. Each bubble represents an event, from yesterday's breakfast, to the building of the great pyramids, to events we call the future simply because we haven't reached them yet.

They all coexist within the same greater whole. What we call the past is the set of bubbles we've moved out of. What we call the future is the set we haven't stepped into.

In this framework, time isn't a line. It's an omni-directional field of possibilities. Our sense of a steady one-way flow arises from biological and cognitive constraints, not from time's true structure.

This matters because it reshapes how we think about distance, movement and cause and effect, not only in physics, but in everyday life. If time is a non-linear dimensional state, and not merely a ruler laid across space, then in principle we shouldn't be slaves to it. It should be something we can interact with, influence and eventually navigate, just as we've learned to navigate oceans and air.

Right now, our tools and understanding keep us locked to a narrow track through this ocean. But if we continue questioning the linear story of time, and begin treating it as a richer, omni-directional realm, we may one day move through it with far more freedom than we currently imagine.

For a deeper look at this Time–Space–Distance paradox, and how non-linear time travel might work in practice, see the chapter *A Concept of Non-Linear Time Travel*.

THE INFINITE ORBITAL UNIVERSE VS THE BIG BANG

Written by Chris Daniels 18th October 2011

I've been encouraged to see cosmology and astrophysics slowly warming to the idea of a rotating, orbital, or whirling universe. It's a promising shift, even if the broader scientific community hasn't yet embraced the larger implications. To my mind, the Universe was never a closed, finite bubble that began at a single explosive point. It's always made more sense to see it as part of an eternal Omniverse, without hard boundaries or a definable beginning.

Back in October 2011, *New Scientist* (Issue 2834) featured Anil Ananthaswamy's report on Professor Michael Longo's work suggesting a spinning universe. The paper proposed that spiral galaxies across vast regions of space appear to favour a particular rotational orientation, hinting at a cosmic-scale angular momentum. That was a refreshing step into territory I'd been advocating for decades. But while the idea of a spinning universe is exciting, it still doesn't go far enough.

I've long argued that the Universe isn't merely rotating. It behaves like a perpetual centrifugal phenomenon, part of a vast, continuous dynamical system embedded within the broader fabric of an eternal Omniverse.

By contrast, I've never aligned with Georges Lemaître's 1931 proposal of the "primeval atom", which Fred Hoyle later labelled the Big Bang. While the Big Bang is often treated as a scientific certainty, at its core it still resembles a creation event: everything from nothing, in a single moment. That's a difficult position to reconcile when there's no evidence that absolute nothingness has ever existed.

It's also worth noting that Lemaître was both a physicist and a Catholic priest. That doesn't invalidate his scientific contribution, but it does highlight the historical context in which the theory emerged. Like many thinkers of his era, he was navigating the boundary between cosmology and metaphysics, at a time when observational data was still sparse.

Once Edwin Hubble's observations of galactic redshift appeared to support an expanding universe, the model gained momentum rapidly, perhaps more rapidly than the evidence alone warranted at that stage.

My own view rejects both the Big Bang and the Steady State model. Instead, I propose countless cosmic universes orbiting autonomously and collectively within an eternal, endless Omniverse. This position isn't born from mysticism, but from observing structural patterns that repeat consistently from the smallest scales to the largest.

Everywhere we look, nature builds through orbital motion.
Electrons orbit atomic nuclei.
Moons orbit planets.
Planets orbit stars.
Stars orbit galactic centres.
Galaxies orbit cluster hubs.
Clusters orbit superclusters.

At every scale, motion is rotational and orbital. To imagine this pattern suddenly stopping at the boundary of our observable universe feels unlikely. Extending the pattern outward leads logically to clusters of universes orbiting greater centres, which themselves orbit even larger hubs, continuing without end throughout the Omniverse.
The more rational question becomes: why *wouldn't* this be the case?

Seen this way, the Universe isn't a creation event. It's a dynamic process operating inside a much larger, ongoing system. Bangs may occur, but they're not creations from nothing. They're energetic consequences of motion, interaction, and friction within the G-Ion fabric of the Omniverse. They're local events, not cosmic beginnings.

This perspective naturally opens into questions about dimensional realms, non-linear time, alternate states of existence, and infinite navigational possibilities across an eternal continuum. In such a framework, time itself isn't a straight line, but a component of an ongoing dynamic field.

Many people tell me they struggle to grasp eternity. They try to picture it, but eternity can't be pictured. If you could picture it, it wouldn't be eternal. What we *can* do, however, is understand it. If existence has no beginning and no end, then something has always been, and always will be. That principle feels far more grounded than the idea of everything erupting from absolute nothing.

Until science fully embraces the possibility of an eternal Omniverse, we'll likely continue hitting theoretical roadblocks. Questions about gravity, time–space structure, dimensional entanglement, and the true scale of reality remain frustrating precisely because they're often approached from a flawed foundational assumption: that the Universe began.

Once we step beyond that assumption, a different landscape forms.
A far bigger one.
A more logical one.
And a more liberating one.

If we accept an eternal Omniverse, we accept an infinite field of possibilities. That's where real progress begins.

THE CENTRE OF THE UNIVERSE AND OMNIVERSE

Thoughts by Christopher Daniels, 1979 – expanded 2025

For decades we've heard the familiar line: *"The Universe is 13.8 billion years old."* It's repeated in documentaries, classrooms, and popular science writing as if it were a settled fact. But this figure has changed repeatedly throughout modern history, and almost always for the same reason: each time we develop a more powerful way to look into space, our estimate of the Universe's size and age expands with it.

Older generations will remember when the Universe was described as nine billion years old. Before that, the number was smaller again. As new telescopes came online, the observable horizon moved outward, and the story changed with it. A deeper view, a brighter instrument, or a more sensitive detector has consistently pushed the limits further, forcing revisions to our estimates.

That alone should make us pause. If our measurements are constrained by how far we can currently see, then our conclusions are constrained by the same boundary. We're attempting to define the scale of existence from inside a very small bubble of observability.

That's why statements framed as *"the Universe is 13.8 billion years old"* are misleading, not because the science is careless, but because the phrasing implies a certainty that doesn't exist. A more accurate statement would be:
"The sphere of our observable universe extends roughly 13.8 billion light-years from Earth."

Even that must be accompanied by the critical caveat that the true extent of the cosmos almost certainly lies far beyond what we can currently calculate or observe.

From this viewpoint, it becomes clear that the age and size of the Universe can't be fixed by the reach of our latest telescope. What we observe isn't the whole. It's only the portion visible to us, through a very narrow keyhole, within a vastly larger construct.

Perspective, not position

Some people still assume that because we can see equally far in every direction, Earth must occupy a central position in the Universe. That's a misunderstanding. It places us at the centre of our *observable sphere*, not at the centre of the cosmos itself.

Sitting in a small boat in the middle of the ocean with no land in sight doesn't mean you're at the centre of Earth's oceans. It only means your view is limited.

Likewise, Earth is a tiny vessel drifting within a cosmic ocean so vast that its boundaries, if they exist at all, remain far beyond our current ability to chart.

The idea of a *centre* here is not a matter of location, but of perspective.

The true centre of the Omniverse

This is where the shift in thinking becomes essential, and where the earlier discussion of *the Illusive Ocularium* connects directly.

In an eternal Omniverse, there is no geometric centre. Eternity has no edges, no boundary, and no outer shell. A centre requires limits, and an infinite system has none. The only meaningful centre is the point from which observation occurs.

Spread your arms and legs to form a human X. Draw a line from your right hand to your left foot, and another from your left hand to your right foot. Where those two lines intersect is, from your perspective, the centre of the Universe. Not metaphorically, but observationally.

If another being, billions of light-years away, performed the same act, the intersecting point of their lines would also represent the centre of the Universe from *their* perspective.

This isn't mystical thinking. It's a straightforward consequence of an infinite system: in a boundless Omniverse, every observer sits at the centre of their own horizon.

This becomes the first conceptual bridge toward understanding the role of the G-Ion. As explored later, the G-Ion is the foundational unit of energy-matter configuration across all dimensional layers.

The G-Ion fabric doesn't merely fill the Omniverse; it constitutes it. It's the lattice through which matter, light, thought, and time register their existence.

Once you understand that perception defines the centre, that the centre shifts with the observer, and that the observer is woven into the universe through the G-Ion fabric, the idea that you occupy the centre of the Eternal Universe isn't arrogance.
It's geometry. It's physics. It's the natural outcome of a boundless system built from a unified substrate.

For a deeper exploration of how the G-Ion fabric establishes structural and energetic symmetry across the Omniverse, see the chapter *Geometric Eternally Metamorphic Ion Theory (G-Ions)*.

Why this matters

If we continue teaching that the Universe began 13.8 billion years ago in a single explosive moment, we confine our thinking to a framework that struggles to accommodate the full scope of reality.

If instead we accept that:
- the Universe is eternal,
- the Omniverse has no boundary,
- the centre exists only relative to an observer,
- and the G-Ion fabric links all points of existence.

Then entire categories of unresolved physics, gravity, entanglement, non-linear time, and dimensional coexistence, begin to open into clearer territory.

In an eternal Omniverse, you aren't located *near* the centre. You occupy a centre defined by perception, just as every other observer does from their own standpoint.

The Universe isn't something happening "out there". It unfolds from wherever observation occurs.

MULTIPLE PERCEPTIONS OF 'ORIGIN'

As discussed earlier, our current technology allows us to see only around thirteen to fourteen billion light-years in any direction. That limitation creates a powerful illusion. Because our view extends outward symmetrically, it's tempting to assume we're positioned near a central point or within reach of a universal origin.

But that impression arises entirely from our vantage point, not from the structure of the Omniverse itself.

What we observe is a sphere, and we sit at the centre of that sphere simply because we're the ones observing from within it. This tells us nothing conclusive about any absolute centre or beginning. It only reflects the limits of light travel and detection.

If the Universe exists within an eternal Omniverse in continual motion, then attempting to trace a straight, linear path back to a single origin becomes unreliable. You can't draw a straight line backward through a dynamic, multi-layered, constantly shifting continuum and expect to arrive at one definitive starting point.

Any observer, anywhere, will naturally identify a different apparent centre because they're measuring from their own moving frame of reference.

This makes the idea of *multiple perceptions of origin* not only plausible, but unavoidable.

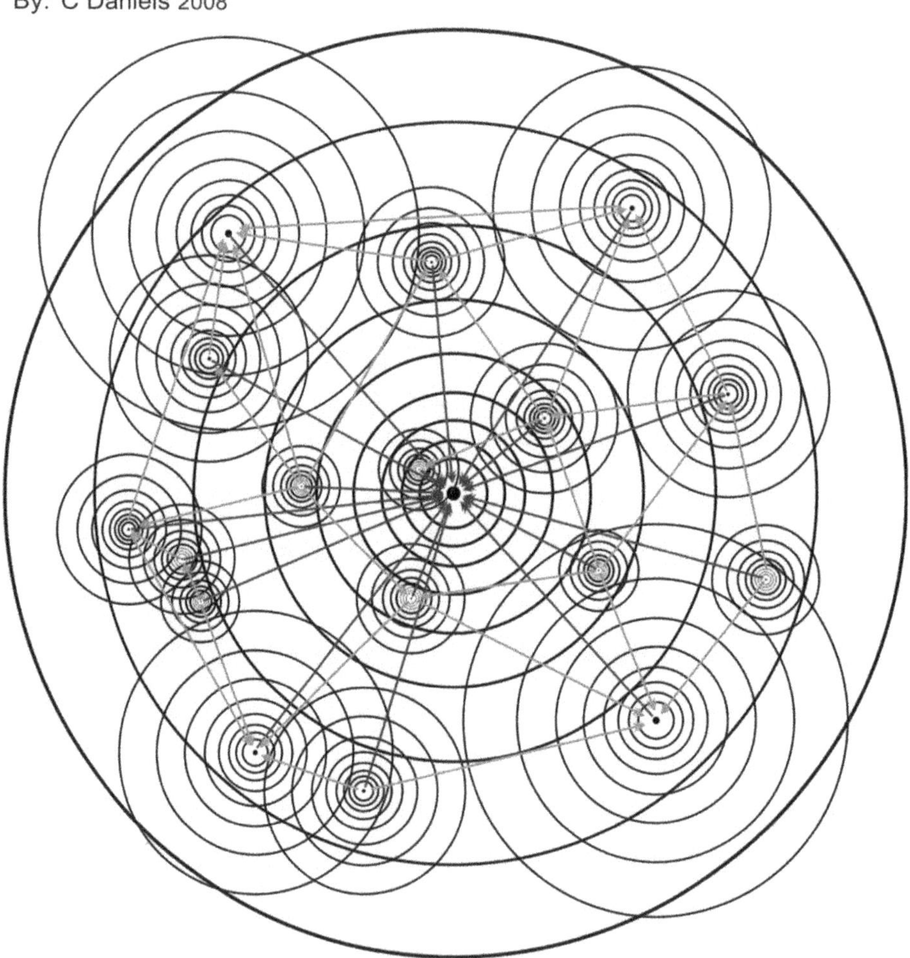

Each point in space and time provides a different frame from which an origin appears to emerge. In reality, there would be countless such perspectives. Every observer, on every world, within every cosmic structure, would infer a different centre based on where they stand within their own horizon.

If time itself is non-linear, as explored earlier, then origins can't be linear either. Observers sample different regions of the same temporal field, not different beginnings.

Taken together, the implication is simple:
There is no single universal origin that all observers would agree upon. There are only origins as perceived from different positions within an eternal, dynamic continuum.

The Omniverse doesn't point back to a central birthplace. It reflects whatever centre the observer's position and perspective define within the endless fabric of space, distance, and time.

SIX PRIMARY DIMENSIONAL REALMS

Before we go any further into non-linear time and the potential for time travel, it helps to establish a clear picture of the six primary dimensional realms we constantly move through, negotiate, and interact with. These realms form the backdrop against which movement, perception, and existence unfold.

In the illustration, the complex symbols are simply expanded versions of the simple ones. They hint at how each realm behaves rather than attempting to define it absolutely. For example, the symbol for Time shows time spreading outward in all directions rather than running along a single straight line. Distance reflects a similar expansion, but with a measured factor. Eternity suggests a never-ending state, while Thought is shown as something that begins at a single point within you and radiates outward indefinitely.

The following symbols are intended to make conceptualising the mechanics of each dimensional realm easier. They're not meant as absolute or authoritative representations, but as visual aids, to help translate abstract behaviour into something our mind can work with.

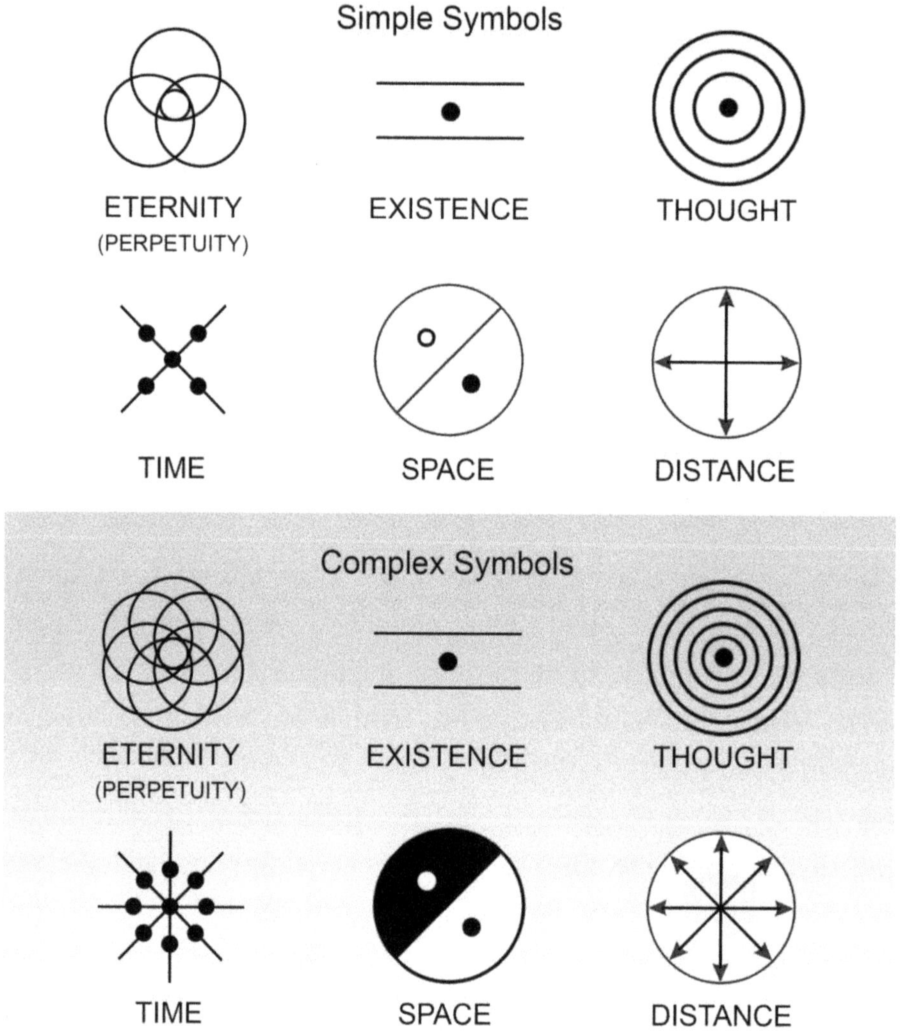

Eternity (Perpetuity)

Eternity isn't a container, a backdrop, or a vast empty space the universe sits inside. It has no edge, no outside, no before and no after. Because it has no boundary, you can't be *in* Eternity in the way you might be in a house or a galaxy.

You, me, every star, every idea, every universe, and every so-called *outside* are already part of Eternity. The symbol reflects endless motion and cyclic interaction, without limit, and with the capacity to intersect time, space, and distance in every direction. Eternity is the prime dimensional realm. If it didn't exist, nothing else could.

Existence

Existence is the realm of being.

The single point in the Existence symbol can represent you, me, an atom, a planet, a galaxy, or any pattern that *is*. It hints that each of us is a tiny expression within a much larger field of being. There may be states that feel like non-existence, but absolute non-existence is, in all likelihood, impossible. The moment you can refer to "nothing", it has already become something within Existence.

Space

Space is any region something can occupy or move within. It can be vast or tightly confined.

The Space symbol suggests directional zones, always relative to an observer. Space only ever makes sense from a point of view.

Distance

Distance is the measure between two points across Space. You can move across it, around it, or through it.

The Distance symbol reflects traversal from an observer's perspective. Distance isn't a thing in itself. It's a relationship between points in Space and Existence.

Time

Time is how we measure movement between points.

In practice, we record how long it takes something to move across, around, or through Space.

The Time symbol reflects that this movement isn't fundamentally linear. From wherever you stand, time radiates outward in all directions.

At a deeper level, time is a dimensional realm that allows events to be ordered, related, revisited, and compared within the wider condition of Eternity. Time appears linear only relative to the observer.

Thought

Thought begins as a point of inception and radiates outward indefinitely. It can be grounded and turned into action, or remain part of a broader flow that connects across time, space, and distance.

The Thought symbol reflects this spreading behaviour. Thought isn't merely an internal sensation. It's an active process that interacts with every other realm.

Why these six realms matter

Some frameworks propose far more than six dimensions. My position is that most additional dimensions, whether philosophical, metaphysical, or scientific, are expressions, conditions, or permutations of these six primary realms.

Once you recognise how interconnected these realms are, non-linear time, multi-dimensional navigation, the perceived boundary of the Universe, and even consciousness itself begin to make far more sense.

The Problem of Containment: Why Eternity cannot be "a place"

This is where many people, including scientists, tend to struggle.

You can't be inside Eternity.
You can't step outside it.
You can't locate its beginning, its border, or its end.
Eternity can't be contained.

A container implies a boundary.
A boundary implies an outside.
An outside implies something larger.

Eternity cannot have an outside.

If something existed beyond Eternity, that *beyond* would itself be part of Eternity. If Eternity had a beginning, something greater must have existed before it, which would mean Eternity wasn't eternal.

This is why the Omniverse isn't an object floating in something else. It isn't a bubble, a region, or a structure. The Omniverse is the expression of Eternity itself.

It's not somewhere we live. It's something we participate in.

That's why every conscious observer stands at the centre of their perceivable Omniverse. A centre can't exist in a boundless thing, but it *does* exist in perspective. Observation creates a centre.

Matter and dimensional states within Eternity

Every phenomenon we describe, from volume and mass to energy, light, and velocity, reduces to behaviours unfolding through combinations of these six framed realms. They aren't additional dimensions. They're expressions of interaction.

For example:
- Volume expresses Space, Distance, and Existence.
- Mass expresses Space and Existence interacting with material density.
- Velocity expresses Space, Distance, Time, and Existence.
- Energy interacts with Time, Space, Distance, Existence, Perpetuity, and Thought.

- Light registers across all six, particularly through wave–particle duality and observational dependence.

Nothing escapes these realms.
Everything depends on them.
And all of them depend on Eternity, not as something surrounding them, but as the condition they arise from.

The Eternal Omniverse: Not a domain, but a condition

If you try to imagine the end of the Omniverse, you reach a contradiction.
If you try to imagine its beginning, the same thing happens.

You can imagine the edge of a local cosmic universe, but not the edge of Eternity.

That's why:
- You can't exit the Omniverse.
- You can't reach its beginning.
- You can't travel to the end of Eternity.

To do any of those things, Eternity would have to be a container. It isn't.

We don't stand outside Eternity.
We don't sit in the middle of it.

We exist as expressions within its condition.

And that's why understanding Eternity isn't optional. It underpins everything that follows, from G-Ion mechanics to non-linear time, traversable space-distance, and the layered structure of matter itself.

The Omniverse simply is.
Eternity simply is.
And we're inseparable from that fact.

A concept of Non-Linear Time Travel

If we could travel at the speed of light, roughly 300,000 kilometres per second, it would take about 4.2 years to reach a terrestrial planet orbiting Proxima Centauri in the Alpha Centauri system. Let's call this planet *Giyora*, meaning *Second Earth*. If we could move at a velocity far beyond light speed, something closer to what I call *thought-speed*, we could arrive almost the instant we turned our attention toward it. But in doing so, we'd find ourselves living 4.2 years in Earth's past by Earth clocks, while standing in Giyora's present. Looking back toward Earth, we'd see it as it was 4.2 years ago, or 4.2 years to come, depending on our interpretive frame.

In June 2014, experimental researchers at the University of Queensland simulated time-travel-like behaviour using photons. Their models suggested that navigating the time–space fabric may be more plausible than once believed. While these experiments don't demonstrate practical time travel, they do reinforce the idea that time may be more flexible than our everyday experience suggests.

The Non-Linear Ocean of Time & the foundation of Time Navigation

It's not only important to understand that time is non-linear, it's vital to recognise time as a single continuum of endless events if we ever hope to interact with it deliberately.

Most people imagine time as a one-way road, where you're either a passenger moving forward or a bystander watching events pass by.

It's a simple picture, but it limits us.

If we stop thinking of time as a line and start imagining it as a vast, omni-directional ocean, where each drop represents a moment or event within a single continuum, we move much closer to what time actually is.

In everyday life, linear time serves us well. It's how we measure movement across space. A day and a year are simply orbital distances travelled by Earth between two points on its journey. But if the dimensional realm we call time didn't exist, our relationship with space and distance would change entirely.

This may sound like science fiction, but so did rockets, satellites, video calls, and atomic theory not long ago. History reminds us that what seems absurd today often becomes ordinary tomorrow. It was once a certainty that the world was flat and everything revolved around us.
If time is non-linear, then everything in time exists simultaneously. Time is not only a measurement, it's also a state of matter-like persistence. That idea can feel confronting, but once the simplicity behind it becomes clear, it may be the only model that truly holds together.

Imagine time as an ocean with an endless horizon in every direction, made up of trillions of tiny bubbles. Each bubble is a complete moment or event. Some contain world-changing events, the extinction of the dinosaurs, the sinking of the Titanic, the invention of the motor car, the founding of nations, or humanity's first step onto an alien world. Others contain smaller moments, like your breakfast this morning or your dinner tomorrow night. They all exist together. We only label them past or future because we've moved away from some and haven't yet entered others.

Reinforcing persistence of moments

This is why the ocean-of-time model is useful. Each bubble persists whether we're present in it or not. There's no delete button in an eternal Omniverse. Moments don't vanish simply because we've moved on. And if moments persist, then in principle, they're navigable.

Somewhere in the Omniverse, there may even be beings whose sensory systems allow them to perceive this ocean directly. What feels impossible to us may be trivial to a mind with broader perception.

People often say we only use ten percent of our brain, which isn't true, but the intuition behind the statement points to something real. Our perceptual limits are biological, not absolute. A million years ago, our ancestors saw far less of the electromagnetic spectrum than we do now. A million years from now, those who evolve from us may see far beyond our current limits, including time itself, as naturally as we see colour and shape.

THOUGHT-SPEED AND THE ILLUSION OF TEMPORAL FRAMES

When we observe the Alpha Centauri system, we're seeing it as it was in the past because light takes time to travel. Proxima Centauri lies about 4.24 light-years from Earth, so the light reaching our eyes left it more than four years ago. If we could move instantaneously to a habitable planet orbiting Proxima Centauri, we'd arrive in that location's present moment, while Earth would immediately become part of our observable past.
This reveals a subtle but important point. What we call time is inseparable from how information moves through space.
Our experience of temporal order arises from delay. Reduce that delay, and the idea of a single, universal, linear timeline begins to break down.

This doesn't mean time is created by distance. It means our perception of time is inseparable from distance and the finite speed at which information propagates. Linear time isn't an absolute backdrop of the Universe. It's a locally constructed framework that emerges from how events are observed and compared. From the perspective of Eternity, time is only linear relative to the observer.

Physical time navigation and its risks

Most people imagine time travel as moving along a line through space-time. But genuine time navigation wouldn't necessarily involve movement in the traditional sense. It would involve entering a state where space and distance no longer apply in the way we're used to.

Without understanding how moment-bubbles overlap, physical attempts would be dangerous. Materialising inside solid matter, unstable environments, or empty space would be a serious risk.

Early experiments, if they ever occur, would almost certainly take place in open space rather than on planetary surfaces, which change too rapidly to be safe.

Navigation itself may prove harder than the jump. Finding your point of origin again could be far more complex than leaving it.

Thought-Speed and Folding Space

Imagine bending the space between two points so they occupy the same location. That's the essence of thought-speed travel. You're not moving faster than light. You're removing the distance between positions.

If Earth were your kitchen and Giyora your bedroom, folding space would mean stepping directly from one to the other. You wouldn't travel. You'd arrive.

Simple in theory. Enormously complex in practice.

But the key lies in understanding the layered nature of matter and the dimensional realms governing space, distance and existence.

Toward the mechanics of time navigation

True time navigation won't come from faster engines. It will come from understanding how matter, space, distance and time are structured at their deepest level.
When we learn how G-Ions organise the layers of matter and how those layers interact with the six primary dimensional realms, bending space becomes a mechanical problem rather than a philosophical one.

Master the G-Ion, and you unlock the navigational mechanics of the Omniverse.

QUANTUM ENTANGLEMENT AND THE 'SPEED OF THOUGHT'

Many readers, particularly those familiar with quantum physics, photonics, or superluminal dynamics, will already know the concept of quantum entanglement.

In simple terms, when two particles interact in certain ways, their properties can become linked. Even when separated by vast distances, a change in one is reflected in the other without any measurable delay. As physicist Michio Kaku has explained, if two electrons interact closely and are then separated, they remain correlated regardless of distance. A change in one is mirrored in the other in a way that appears instantaneous, faster than light could travel between them.

That apparent immediacy is what I refer to as *thought speed.*

This doesn't mean literal thought is pushing particles around. It refers to a class of interaction that operates outside the familiar constraints of light-bound signalling. Entangled particles don't communicate by sending energy or information through space in the conventional sense. Instead, their relationship exists as a single quantum system, even when spatially separated.

This behaviour troubled Albert Einstein. His theory of relativity established light speed as the upper limit for the transmission of information, and entanglement seemed to violate that rule. In 1935, Einstein, Boris Podolsky and Nathan Rosen attempted to demonstrate what they believed was an incompleteness in quantum mechanics.
Instead, their work gave rise to what is now known as the EPR paradox, a formal description of precisely the kind of non-local behaviour quantum experiments later confirmed.

Einstein famously described this as "spooky action at a distance".

Ironically, in trying to undermine quantum theory, the EPR paper helped reveal that reality may permit interactions that are not constrained by distance in the way classical physics assumes.

Quantum entanglement doesn't break causality, but it does force us to accept that correlation can exist without traversal. That distinction matters. It hints at an underlying structure of reality where connection is not always mediated by motion.

Every major advance in science began as an imagined framework. Rather than dismissing this behaviour as an anomaly, it may be more productive to explore what it's telling us about the deeper fabric of the universe. Entanglement may represent our first clear glimpse of a regime where influence and relationship operate at scales and speeds that make light-based interaction look almost static by comparison.

If the universe already supports instantaneous correlation across vast distances, it raises a deeper question. Why should our future methods of travel remain confined to mechanical propulsion alone, when reality itself appears to operate on principles far more efficient?

This is where the concept of thought speed becomes useful. Not as a claim, but as a guiding idea that encourages us to rethink what advanced navigation might eventually involve!

THOUGHT SPEED VS NUCLEAR, FUSION AND WARP-BASED PROPULSION

Today, enormous effort and funding are being directed toward advanced interstellar propulsion concepts. Nuclear fusion engines, antimatter containment, artificial gravity systems, warp field experiments, and solar or laser-driven light sails are all under active investigation.

Since 2013, NASA has hosted formal warp field research programs at the Johnson Space Center, led by physicist Harold G. "Sonny" White. These studies explore whether space-time itself can be engineered for travel, rather than treated as a passive background. More recently, related research has moved into classified domains. In early 2025, it was revealed that discoveries with potential relevance to warp mechanics are under active investigation within the U.S. Defense Advanced Research Projects Agency.

Warp drive concepts originate from solutions to Einstein's field equations, most notably the Alcubierre metric proposed in 1994. In this framework, space-time expands in front of a craft and contracts behind it, allowing effective faster-than-light travel without locally exceeding light speed. The craft doesn't move through space in the usual sense. Space moves around it.

The primary obstacle has always been energy. Early models required astronomical quantities of exotic matter with negative energy density. White's work significantly reduced those theoretical requirements, showing that optimised warp geometries could, in principle, bring the energy demands closer to feasible scales.

More recent investigations have explored whether quantum-scale effects, such as those associated with the Casimir effect, might support microscopic warp-like distortions.

Even if these approaches prove viable, they remain forms of distance-based navigation. They shorten paths, compress space, or alter geometry, but they still depend on energy expenditure and transit mechanics.

Thought speed implies something fundamentally different.

Quantum entanglement already demonstrates that correlation can occur without traversal. No known propulsion system, warp-based or otherwise, replicates this behaviour. Warp drive reshapes space. Thought speed questions whether space must be crossed at all.

This distinction is crucial. Warp drive represents an advanced extension of physical navigation. Thought speed points toward a deeper relationship between information, cognition and the structure of reality, where location becomes secondary to state.

Fusion engines, solar sails and warp fields are not dead ends. They're stepping stones. Just as sailing ships led to steam, steam to combustion, and combustion to nuclear power, space propulsion is evolving toward increasingly abstract interactions with the fabric of space itself.

Each step teaches us more about what space, time and motion truly are. And each step brings us closer to a far larger leap: not moving faster through space, but learning how to step outside its constraints altogether.

I've long suggested that humanity may one day develop pulsar-based navigation systems, using pulsars as interstellar beacons, reference points and energy anchors. Such systems could form the backbone of an early cosmic map.

Warp drive may be the cathedral under construction. Thought speed may be the moment we realise we no longer need walls, corridors or roads at all.

THOUGHT AND THE PARTICLE WAVE THEORY

Before we go any further, we need to entertain a simple but important idea. Thought, as one of the six primary dimensional realms, may also be a physical property. If that's true, then it should, in principle, be measurable in the same way we measure the behaviour of light or sound. We may not yet have the instruments to do it, but that doesn't mean the property isn't there.

It's worth making one point clear before continuing. Earlier in the book, we explored the idea that thought might not rely on velocity at all. Thought could operate as a form of spatial shift, similar to bending space or collapsing distance, meaning it may reach a destination without travelling through the space between two points. That idea still stands.

Here, however, we're also exploring another valid possibility. Thought may not only behave dimensionally, it may also exist as a form of particle-matter, much like light and sound do. If that's the case, then thought could possess a measurable velocity of its own, potentially far beyond the speed of light, carried by particles or quanta we've not yet identified.

These two possibilities are not in conflict. Until we can measure thought directly, both models deserve consideration. Thought may turn out to be a dual-state phenomenon, behaving like a physical particle under some conditions and like a non-local dimensional field under others, much as photons exhibit both wave and particle behaviour. In an Omniverse built on layered matter and multi-dimensional states, this kind of dual or even multi-modal behaviour shouldn't surprise us.

If you're already familiar with the principles behind G-Ions, you'll recognise how this fits within the behaviour of transfer particle waves across the G-Ion ocean. In that framework, thought behaves as one of the unknown quanta moving through the fabric of existence, carried across the G-Ion medium with direction, intensity and velocity.

Put simply, thought might not be only an internal experience. It may be something that moves from one point to another, in much the same way a photon moves from star to eye. The time it takes for a unit of thought to reach its destination would then define its velocity. That would be the speed of thought.

If you're not yet familiar with G-Ion theory, there's a full explanation later in the chapter titled *Geometric Eternally Metamorphic Ion theory (G-Ions)*. You don't need it yet. For now, it's enough to hold the idea that the fabric of reality may carry thought in the same way it carries light. The difficulty lies not in the concept, but in measurement.

At present, the speed of thought sits in an indeterminate category of physics. We haven't yet developed tools capable of detecting the motion of thought-carrying particles, or whatever unknown quanta may underpin them. This creates a familiar loop. We can't declare thought to be a measurable physical property until we measure it, and we can't measure it until we develop instruments capable of detecting it. All we can do for now is acknowledge the probability.

History suggests that this position is not unusual. Before sound waves were understood, the idea that sound had structure was uncertain. Before light was quantified, its dual wave–particle nature was speculative. Before atoms were observed directly, they existed only as philosophical constructs.
In every case, technology eventually caught up with imagination.

Given that pattern, should we really dismiss the possibility that thought might also be a physical phenomenon, travelling at extreme velocity, perhaps even far beyond the speed of light, along directed or diffuse trajectories we simply haven't learned to detect yet? History suggests we shouldn't dismiss the possibility. Every now-known physical property began as an unknown quantity.

Some emerging scientific discussions quietly point in this direction. Questions around simulated or virtual reality models have led some researchers to revisit the foundations of relativity itself. Claims made decades ago that Einstein's equations might be incomplete were dismissed at the time, only to resurface when quantum mechanics exposed gaps in classical assumptions. Quantum theory continues to reveal behaviours that sit uncomfortably inside the boundaries of earlier frameworks.

Does this validate the speed-of-thought concept? Not yet. But it keeps the door open. It invites us to treat thought as a candidate for physical investigation, rather than confining it to biology or metaphysics alone.

If thought does propagate, and if it does so at extreme velocity, then eventually we'll need a symbol for it in physics and mathematics. For now, I use *ts* (thought speed), *sot* (speed of thought), or more formally *Thv* (thought velocity) when discussing the concept in applied formulations.

We're not there yet. But the first step is recognising that thought may be part of the physical story of the universe, not an exception to it.

Hypothetical Considerations

Let's consider a hypothetical scenario in which thought has already been detected as a physical property, confirmed to behave as matter, and shown to enter a state of momentum. In this imagined but internally consistent framework, our technology is advanced enough to detect thought travelling with velocity and to measure that velocity as dramatically exceeding the speed of light.

For the sake of this exercise, assume we've discovered that thought, when in momentum, travels faster than light by a factor of four.

Let the symbol **th** represent thought, and **Thv** represent the speed, or velocity, of thought.

The basic relation can then be expressed as:
Thv = c x 4

Where:
c = speed of light = 299,792 kilometres per second
    (or, 186,282 miles per second)

Common scientific approximation: $c \approx 3 \times 10^5$ km/s

This establishes a simple, testable starting point for treating thought as a measurable physical property, without making any claims about mechanism, origin, or application.
It's not presented as proof, but as a structured hypothesis, framed in the same way early models of sound, light, and atomic behaviour were once framed, long before their direct measurement became possible.

THROUGH TIME AT THE 'SPEED OF THOUGHT'

At this hypothetical velocity of **1,199,168 kilometres per second**, it would be possible to travel from the point where we observe a beam of light back toward its origin and reach that origin *before the light ever arrived.*

To make this easier to visualise, imagine you're observing a star whose light takes **ten years** to reach your eyes. If you then travelled back along that same path at a speed four times faster than light, you'd arrive at the star in **two and a half years**. From Earth's frame of reference, you would reach the star *seven and a half years before the light you observed was emitted.* Depending on relative motion and reference frame, you could even arrive at a point before key events in that star's history had occurred.

Now consider a far more plausible extreme.

Thought speed may exceed light speed not by a factor of four, but by a factor of 10^{11} which is a one followed by eleven zeros.

That gives:
Thv $= c \times 10^{11}$

Since:
$c \approx 3 \times 10^{5}$ kilometres per second

Then:
Thv $\approx 3 \times 10^{5} \times 10^{11}$
Thv $\approx 3 \times 10^{16}$ km/s

Which is:
Thirty quadrillion kilometres per second
(30,000,000,000,000,000 km/s)

Because the speed of light already includes five powers of ten, multiplying it by 10^{11} simply adds those powers together, resulting in a velocity on the order of 10^{16} **kilometres per second**.

At that scale, any star ten light-years away becomes reachable in what would, for all practical purposes, feel like an instant.

Yet the more likely truth is that thought speed is not merely extreme, but *immeasurable*. In that case, velocity itself becomes the wrong language. Distance ceases to matter at all. This figure is not proposed as a literal speed through space, but as a conceptual contrast, illustrating how radically non-local information transfer would differ from any propulsion-based motion, no matter how advanced.

At that point, the core challenge shifts away from speed and toward **navigation**. If distance collapses, orientation becomes the real problem. We'd still need a defined operational framework to negotiate movement through the time–space ocean, just as we require reference systems and coordinates to navigate physical space.

CALCULATION OF THOUGHT-SPEED OVER 10 LIGHT-YEARS:

For context, **Proxima Centauri** lies approximately **4.2 light-years** from Earth. Several known stars fall close to the ten light-year range, including:

- Ross 154 at 9.4 light-years
- Ross 248 at 10.3 light-years
- Epsilon Eridani at **10.7 light-years**

For simplicity, let's imagine an artificial reference star located at exactly **10 light-years** from Earth. We'll call it **Alpha Australis**.

A light-year is the distance light travels in one year.

Ten years corresponds to approximately:
10 years $\approx$ 315,360,000 seconds
(3.1536×10^8 seconds)

Distance travelled by light in ten years:
Distance = time x c
$$= (3.1536 \times 10^8 \text{ s}) \times (3 \times 10^5 \text{ km/s})$$
$$\approx 9.4608 \times 10^{13} \text{ kilometres}$$

From the earlier thought-speed estimate:
Thv = 3×10^{16} kilometres per second

Travel time at thought speed:
Time = Distance / Thv
$$= (9.4608 \times 10^{13} \text{ km}) / (3 \times 10^{16} \text{ km/s})$$
$$\approx 0.0031536 \text{ seconds}$$

Which means:

- **Less than 0.005 seconds** to reach Alpha Australis
- **Less than 0.01 seconds** to travel there and return (without pause)

If Alpha Australis had only just been created, arrival at this velocity would occur *before* its formation event from the observer's frame of reference. One could, in principle, observe the moment prior to its first light radiating outward, or even encounter the conditions that led to its creation.

This immediately raises deeper questions.

If arrival precedes creation from your perspective, does that imply the possibility of influence?

Could intervention alter the outcome?

Or would such a visit simply become part of the causal sequence that leads to the star's existence?

These are not rhetorical flourishes. They're the unavoidable paradoxes that emerge once non-linear time, extreme non-local interaction, and observer-relative frames are taken seriously. They also underline why understanding the mechanics of thought, perception, dimensional layering, and G-Ion behaviour must come *before* any meaningful discussion of temporal navigation.

Time travel, if it exists at all, is not a problem of speed.

It's a problem of structure.

THOUGHT AS A PHYSICAL PROPERTY & MECHANICS OF THOUGHT TRAVEL.

Before we go deeper into how thought may move through time and space, we need to address one key idea: if thought can interact with the universe in any measurable way, then it must possess physicality. It must be a property of the Omniverse itself, not merely a private internal experience. This doesn't mean thought behaves like ordinary matter or conventional energy. In fact, it appears to behave very differently. But strangeness has never disqualified a phenomenon from physics.

What we're exploring here is the possibility that thought functions as a physical quantity with measurable properties, capable of momentum, direction, and interaction. This doesn't abolish existing physical laws. It extends them, in much the same way quantum mechanics extended classical physics without replacing it.

Once we accept this premise, a natural question arises: how could something as subtle as thought operate at velocities that appear to exceed the speed of light by enormous margins? The answer may lie not in how fast thought moves, but in *how* it moves. Thought is unlikely to propagate as a single, isolated particle travelling through space. Instead, it appears to operate through a transfer mechanism conceptually closer to quantum entanglement, but operating at a deeper and more general level.

The Particle Wave Transfer Mechanism

Thought appears to propagate through what I call **Particle Wave Transfer**.

Rather than a discrete unit of thought travelling across space like a photon, the informational structure of the thought is transferred across a vast, interconnected field of G-Ions.

Each G-Ion functions as both carrier and translator, relaying the complete informational pattern of the originating thought to the next G-Ion in sequence. What arrives at the destination is not a degraded signal or approximation, but a precise informational duplicate, indistinguishable from the original thought that initiated the transfer.

From our perspective, this appears instantaneous.

A thought transference G-Ion's cloning path

The information doesn't traverse distance in any classical sense. It is re-expressed across the fabric itself. If this sounds reminiscent of quantum entanglement, that's no accident. Entanglement is a visible hint of the behaviour. The G-Ion framework describes the substrate that makes such behaviour possible across all scales, not just in isolated quantum systems. Unlike photons, which are constrained by space-time geometry, G-Ions are not limited by conventional spatial boundaries. They permeate all matter, all layers, and all dimensional realms. As a result, they can negotiate obstacles, traverse dense material, and adapt geometric orientation without requiring a continuous linear path.

In this sense, thought doesn't move *around* matter. It moves *through* it.

This also explains why thought doesn't require a defined route. If the destination lies on the far side of a planet, thought doesn't travel around the planet. It propagates through the same underlying fabric that constitutes the planet itself. G-Ions are not separate from matter. Matter is one of their organised expressions.

The Scale Problem and Why Thought Goes Unnoticed

Our difficulty in detecting these processes isn't evidence against them. It's a consequence of scale.

The interactions involved occur far below the threshold of current instrumentation, operating at atomic, femtometre, attometre, and potentially zeptometre scales. A zeptometre is one-millionth of one-billionth of a metre. Processes unfolding at this level will appear instantaneous, non-local, and invisible to us until our measurement technologies evolve accordingly.

Humanity once struggled to detect atoms, electricity, and electromagnetic fields. Today, they form the foundation of modern science and technology.
There's no compelling reason to assume thought must be exempt from this pattern.

The barrier isn't impossibility. It's resolution.

Once our tools catch up to the scale at which thought operates, what currently feels intangible may become measurable, describable, and eventually navigable. The leap required here is conceptual, not physical.

The Physicality of Thought travel

If thought behaves as a physical property, and if it uses the G-Ion fabric to transfer information at extreme velocity, or even outside conventional notions of velocity altogether, then the implications for time navigation are profound.

Thought may be capable of interacting across both forward and backward frames of time. Not metaphorically, but physically, as an informational process embedded within the same fabric that structures matter, space, and time itself. We may already experience faint echoes of this capacity when we recall memories, anticipate future events, or sense outcomes we cannot logically justify. These experiences may represent low-resolution manifestations of a process we haven't yet learned to access deliberately or with precision.

Consider the following scenario.

You project a structured thought toward a location ten light-years away. Because the transfer occurs outside the limitations of light-speed propagation, the informational structure of that thought reaches its destination before the star's light ever reaches Earth. If information is then returned through the same mechanism, you are interacting with data originating from a different temporal frame relative to your local reference point.

In principle, this allows for interaction across both past-aligned and future-aligned temporal states, not by violating causality, but by operating within a non-linear temporal structure where events coexist within a broader continuum.

This raises important and challenging questions.

If information can be accessed from a moment prior to your biological existence, does that constitute participation in an event beyond your physical lifespan? And if information can be retrieved from a future temporal frame, does that represent interaction with a state that has not yet been locally realised?

These questions don't belong to science fiction. They emerge naturally once we treat thought as a physical process rather than a purely internal phenomenon. The mechanism doesn't rely on breaking physics, but on extending it into domains that current models don't yet describe.

When thought can be measured as a physical property, when its interaction with the G-Ion fabric can be quantified, and when the mechanics of information transfer across space–time are better understood, thought-based navigation ceases to be speculative.
It becomes a logical consequence of an expanded physical framework.

What feels like a leap today may one day be recognised as one of the most intuitive extensions of physics we ever overlooked.

CONSIDERING REAL POTENTIALS

Now let's do something useful. Let's temporarily set aside the scientific impossibilities we instinctively attach to these ideas. Not as dismissal, but as an exercise in exploring what becomes plausible once we loosen the constraints imposed by current technology and incomplete understanding.

The first potential is both simple and profound. Thought may be a physical property that can be measured. If thought is physical, it can interact with momentum, engage in motion, and operate at velocities far beyond the speed of light. At the same time, thought exhibits behaviours unlike any familiar particle. It shows quantum-like characteristics, including apparent non-locality, yet doesn't sit comfortably within existing quantum models. That places it within an unexplored region of physics and geometry, a frontier that remains largely unmapped.

If thought travels through the Omniverse using its own transfer mechanism, then its effective traversal of mass distances may appear instantaneous. From our limited frame of reference, distances we currently describe as vast or unreachable could become negligible. A shift of tens of thousands of light-years might, functionally, be no more significant than moving a few metres across a room. What appears insurmountable at the human scale may be trivial at the zeptoscopic scale of the G-Ion fabric.

We already accept that thought arises from bio-mental processes involving conscious, subconscious, and deeper cognitive activity. If that process produces a physical property, then we remain physically connected to the thoughts we generate. And if that property can propagate, then some component of us propagates with it, carried along the pathways created by our own cognitive output.

This doesn't imply that the human body moves through time. But it does suggest that one component of our total construct may be capable of doing so.

That possibility opens a new theoretical door, one that leads to an intriguing question: are we already engaging in a rudimentary form of time interaction without realising it?

Which brings us to one of the most under-examined experiences in human life.

PERHAPS WE DO 'TIME TRAVEL'

Before diving deeper into quantum and G-Ion mechanics, it's worth examining a softer, more familiar phenomenon. For thousands of years, dreaming has been surrounded by myth, symbolism, and speculation. Modern neuroscience has clarified much about sleep cycles and REM activity, yet some dream experiences still resist straightforward explanation.

Across cultures and historical periods, there are persistent reports of people dreaming events that later occur, dreaming places they've never visited yet later recognise, or dreaming interactions later corroborated by external discovery. Research suggests that a large proportion of people experience dreams vivid enough to feel indistinguishable from waking reality.
If thought is a physical property capable of propagating through the G-Ion fabric, then some dreams may not be internal constructions at all. They may be receptions. Moments where information is accessed from other points in time, space, or distance, received when conscious filtering is reduced and the subconscious becomes more permissive.

In metaphysical traditions, this is often described as out-of-body experience or astral travel. From a scientific standpoint, it may represent cognitive interaction with the same thought-wave transfer mechanisms discussed earlier. If thought can traverse time and distance, then dreams may be moments when this process occurs without deliberate control.

History gives us perspective here. Invisible forces travelling through space were once dismissed as fantasy. Radio waves would have sounded like sorcery in the sixteenth century. Particle–wave duality would have seemed absurd to pre-modern thinkers. Video calls would have been indistinguishable from magic.

Seen in that light, it's not unreasonable to consider that some dreams may represent genuine physical interactions mediated by a thought-wave field. If so, dreaming may be humanity's oldest and most common interface with non-linear time.

BEYOND THE REALM OF SCIENCE

If thought is a physical property capable of motion through the G-Ion fabric, then its implications extend beyond physics alone. They reach into domains traditionally labelled metaphysical or mystical.

Experiences such as intuition, premonition, déjà vu, prophetic impressions, or even reported past-life memories may all reflect interactions with physics we don't yet fully understand.

For instance, what's often described as the past-life effect may not involve reincarnation at all. It could arise from interaction with other temporal frames through thought-wave processes. An individual might access memories associated with another life trajectory, integrate those impressions, and interpret them as personal identity.

Human psychology already shows how readily identity adapts. People entering new cultural environments, military units, communal systems, or isolated contexts often internalise new roles rapidly. In more extreme cases, dissociative identity phenomena demonstrate how convincingly the brain can generate complete, lived identities. When viewed through the lens of time-based thought interaction, the absorption of impressions from another life-frame becomes far more comprehensible.

None of this asserts fact. It simply acknowledges possibilities that align with known psychological behaviour and the physical framework being explored.

On the Simultaneous Emergence of Ideas

There's another phenomenon worth noting here, one that's often dismissed as coincidence but appears far too structured to ignore.

Throughout history, the same ideas have emerged independently in different minds, at different locations, often within the same narrow window of time.
Entire scientific theories, technological solutions, artistic narratives and philosophical frameworks have appeared almost simultaneously across the globe, without any direct communication between the individuals involved.

This isn't speculation. It's a recognised pattern. And the more complex the idea, the harder it becomes to explain purely through chance.

If thought were entirely local, isolated inside individual brains, this kind of synchronised emergence would be statistically unlikely. But if thought is a physical process interacting with an eternal informational fabric, the phenomenon becomes far easier to understand. In that case, ideas don't originate *in* people so much as they are *accessed* by them.

Much like moments in the ocean of time, thought may already exist as part of the Omniverse's perpetual structure. An idea doesn't begin when someone thinks it. It becomes active when a mind reaches the conditions required to engage with it. Each idea remains original in its own domain, even if more than one individual arrives at it independently, because originality lies in the act of access, interpretation and application, not in exclusive ownership of the concept itself.

A Conclusion without a conclusion

So does the likelihood that thought is a physical property, capable of propagating at extreme velocities across phenomenal distances, imply that time or cross-dimensional interaction is possible?

I believe it strongly suggests that it is, and that it may already be occurring.

But evidence must come from those who follow, equipped with tools we don't yet possess. Their role will be to test, refine, or replace these ideas with models that stand up to rigorous scrutiny.

All meaningful progress begins with a question. And sometimes the question has to come before the instruments capable of answering it.

A quick recap on Fringe Science

Fringe science isn't pseudoscience. It's the frontier between the known and the not-yet-known. It represents ideas that step outside orthodox frameworks while still showing enough conceptual structure to justify exploration. Some of these ideas eventually become mainstream once evidence and technology catch up. Others remain incomplete, but still valuable for what they reveal and the questions they provoke.

Fringe science is the testing ground where new theories challenge old assumptions. It's where imagination is allowed to operate alongside method, and where uncomfortable questions are not dismissed simply because they don't yet fit existing models.

Quantum Physics on the Fringe

Quantum physics itself was fringe not long ago. Today, it underpins some of the most advanced technologies we possess. Yet many aspects of quantum behaviour still resist comfortable integration into classical physics. Particles behave in ways that defy intuitive expectation. Systems change when observed. Entangled pairs act as though distance is irrelevant.

Certain features of the quantum field display organisation and responsiveness that challenge purely mechanical descriptions.

This isn't mysticism. It's simply where the evidence continues to point.

One direction quantum physics increasingly gestures toward is the idea of a universal simultaneity. Not a point in space, but a condition in which all moments coexist. Every particle, every interaction, every possibility. This aligns closely with what we've explored throughout this book: time as an ocean rather than a line. A non-linear realm where everything exists together, accessed through perspective rather than sequence.

This framework doesn't answer questions about life beyond death or consciousness beyond biology. But it does open the door to investigating them with intellectual honesty instead of superstition.

Quantum physics is real science, even when it operates at the edge of understanding. And if we remain disciplined, holding imagination and evidence to the same standard, the fringe becomes the workshop where tomorrow's discoveries take shape.

As always, it comes back to the guiding philosophy of the true explorer: "Always consider the possibility of anything, while looking for the facts and probability of everything."

And one final reminder worth carrying forward:

"Everything you can imagine exists in the Omniverse, but the Omniverse is filled with things you will never imagine."

(Revised James Cole quote, 1995)

Geometric Eternally Metamorphic Ion theory (G-Ions)

Hypothesis, and developments study 1986 – 2019

Definition of 'Theory'

A theory is simply a structured idea. It's a system of principles intended to explain a phenomenon, or at minimum to offer a coherent way of thinking about complex behaviour. A theory does not need to be proven to be valuable. Some theories guide experimentation. Others guide imagination and exploration.
The Gem-Ion theory functions as both.

Why G-Ion or Gem-Ion

The name *Gem-Ion*, or *G-Ion* for short, reflects the proposed behaviour and appearance of these entities. They are envisioned as geometric structures capable of shifting form in ways that resemble crystalline or faceted configurations, with multidimensional faces and vertices.

The term captures both the geometric character of the particle and its capacity to metamorphose while retaining an underlying structural coherence.

When visualised, G-Ions appear almost jewel-like, exhibiting symmetry and facet behaviour, yet they are fundamentally amorphous. Their geometry is not fixed. It is adaptive, capable of shifting instantly to suit local conditions, interaction demands, or information-transfer requirements.

This capacity to change form without losing structural identity provides a useful way to imagine how G-Ions could mediate matter, energy, and information, including the rapid imprinting and cloning processes discussed earlier in the chapter on *Thought as a Physical Property*, and explored in greater depth here.

This imagery turns out to be more than illustrative. It offers a functional way to understand how such entities might build, structure, and influence physical universes, not through static form, but through continuous geometric response to need, context, and connection.

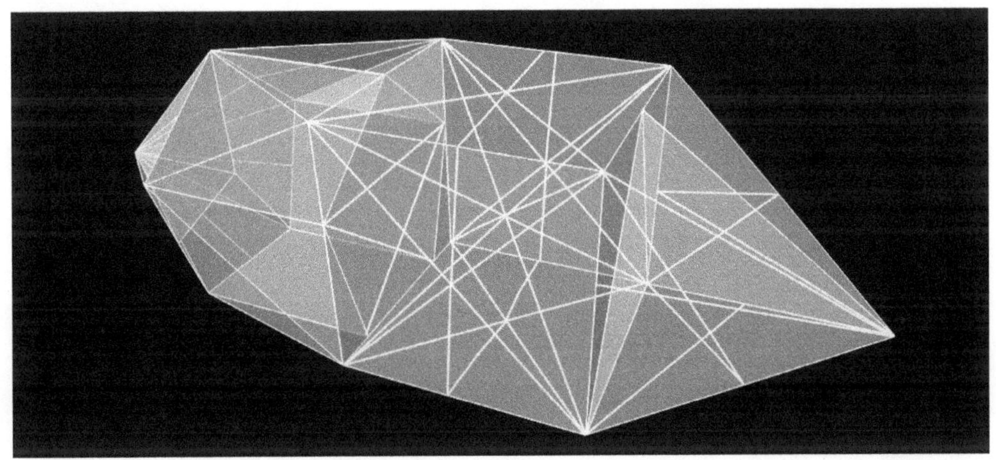

(Example of a single G-Ion cell structure, illustrating one of its endlessly adaptive amorphous geometries)

G-Ion: Mechanics, Modality, Law and Behaviour

The G-Ion theory proposes that beneath all known matter, forces, and fields exists a deeper fabric composed of perpetually dynamic, endlessly shifting entities. These entities are not confined by the dimensional constraints of quantum objects as they're currently defined.
They may operate at scales smaller than any presently measurable particle, while also participating in collective behaviours that extend across cosmic structure.

Rather than existing as isolated units, G-Ions form a unified, continuous ocean. Everything in existence, from ordinary matter to antimatter, dark matter, and dark energy, is interpreted as emergent behaviour within this ocean.

In this model, gravity is not a force acting *upon* G-Ions. It's a behaviour *of* the G-Ion fabric itself. Gravity is not imposed from outside the system. It emerges from internal motion, pressure differentials, and structural response within the ocean. G-Ions cycle continuously through expansion and compression. They're capable of super-dense clustering, forming the conditions required for the hardest known materials and the most rigid structures. In the opposite regime, they can become more fluid and subtle than vapour or plasma, creating extremely low-viscosity states.

These transitions occur at speeds and scales far below current measurement thresholds, often within intervals far shorter than nanoseconds.

G-Ions also carry and generate energy. Their intersecting vertices produce electromagnetic and sub-electromagnetic interactions that release radiant energy into the Omniverse. This includes both visible energy and what is presently classified as dark energy.

In this sense, G-Ions function as the structural and energetic basis of physical reality. They give matter its volume, shape, density, and interactive behaviour. They contribute to the emergence of forces and the organisation of structure. Nothing within the observable universe exists outside the influence of G-Ion dynamics.

Where the theory becomes especially distinctive is in how G-Ions communicate.

When a G-Ion undergoes positional change, it does not travel through space in a linear sense. Instead, it transfers a complete geometric and informational imprint of itself to the adjacent G-Ion within the contiguous fabric. That neighbouring G-Ion becomes the next instantiation of the original state.

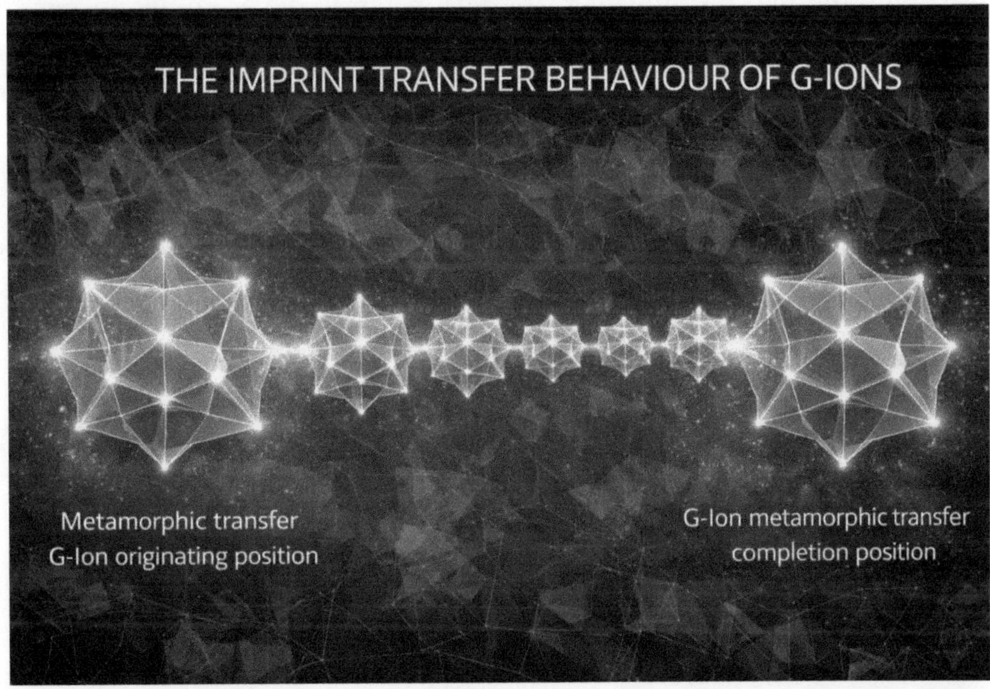

This creates a chain of metamorphic replication rather than particle translation. This process is referred to as **metamorphic imprint transfer behaviour**.

The originating G-Ion remains where it is. What propagates is a perfect unfolding of its informational structure.

This aligns with the transfer-wave behaviour discussed in earlier chapters. The effective propagation speed of this process is not limited by the speed of light. It more closely resembles what has been described as *Thought Speed*, as imprint transfer occurs across the fabric without traversing distance in a classical sense.

A useful analogy is to imagine a vast ball-pit filled with perfectly fitted spheres. If all spheres were static, nothing would move. Introduce entities that deform, shift, and apply pressure, and motion propagates throughout the system. Multiply this by continuous geometric adaptation, and the result is a dynamic, self-organising ocean.

Unlike the ball-pit, however, the G-Ion fabric never comes to rest. No region becomes motionless. The universe persists because the underlying fabric is in constant transformation.

Energy generation arises from continual collision, repulsion, and magnetic tension at intersecting G-Ion vertices. The interplay between compression and displacement produces high-density energetic states, which can become unstable under specific conditions. This offers a conceptual framework for phenomena ranging from nuclear processes and stellar lifecycles to plasma behaviour and the gravitational environments surrounding black holes.

Because G-Ions are energetic as well as structural, they naturally organise with others of compatible energetic signatures.

This behaviour underpins the formation of elemental groupings that later express as atoms and molecules.

The crucial distinction between G-Ions and known subatomic particles is continuity. G-Ions are never isolated. There are no gaps between them. They form an unbroken fabric.

When one region of the fabric is displaced, neighbouring regions must respond. Displacement triggers reconfiguration, and reconfiguration propagates further displacement. From this behaviour, gravitational effects emerge naturally.

Gravity, in this framework, is not a force acting on matter. It is the collective response of matter's underlying fabric to internal motion.

This insight forms the foundation for the gravitational and magnetic behaviours explored later in this chapter, and for the technological implications discussed in subsequent sections.

CONSTRUCTION AND STRUCTURE OF G-IONS

So far, we've treated G-Ions as the underlying substance of everything. To make that idea more concrete, it helps to zoom in and look at how they're structured.

In this theory, a G-Ion is a specific geometric construct made of connected cells that can group into larger collectives.

Those collectives can then form new, autonomous structures, all while remaining locked into a mathematically coherent, ever-shifting jigsaw that fills the Omniverse without gaps.

You can think of the hierarchy roughly like this:

- G-Ion cell (GI cell)
  A single section or face of a G-Ion. One tile within the frame.

- Single G-Ion

  A three-dimensional geometric structure formed by many GI cells sharing the same internal space. This is one complete G-Ion unit.

- **Cluster**

  A dense collection of single G-Ions, including Active, Non-active, and Re-active Unstable G-Ions. From the outside, it may look messy or unstructured, but internally it behaves as a cooperative system.

- **Element**

  A compact, enclosed collective of G-Ion clusters forming a larger geometric structure. At familiar scales, these Elements may manifest as quarks, subatomic particles, or other fundamental units.

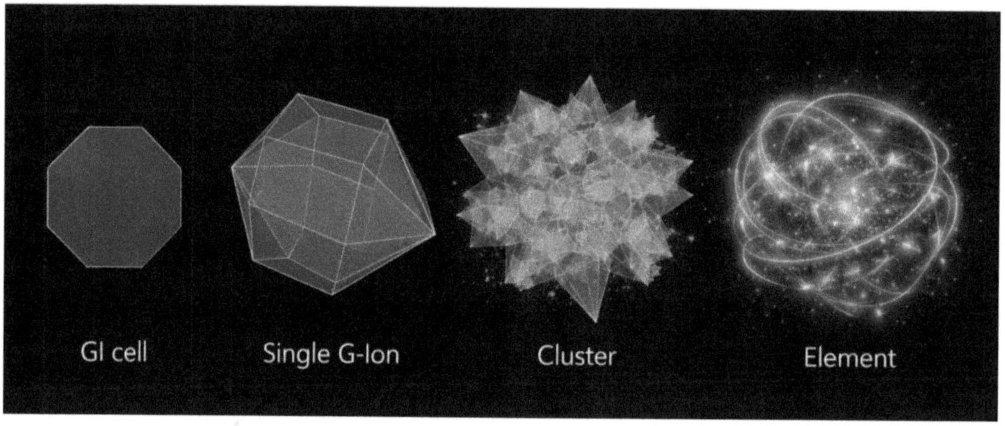

From the outside, this can easily look like "just another way of labelling known particles". The key difference is that, in the G-Ion framework, those familiar particles are *consequences* of the G-Ion fabric, not the base layer itself. G-Ions sit beneath, acting as the true carriers of volume, connectivity, and energy.

Several important points follow from this:

- G-Ions do not possess solid mass in the conventional sense.
  Their form is defined by the strings or edges connecting their vertices. Because those strings belong to a single continuous fabric, a G-Ion behaves less like a tiny marble and more like a patterned region of tension and vibration.
- What appears to be empty space within a G-Ion frame is not empty.
  It's intense energy generated by the perpetual motion of the fabric. Depending on state, a G-Ion may appear solid, non-solid, luminous, dark, or somewhere in between. At large scales, this shifting behaviour contributes to what we currently classify as dark matter or dark energy.

- Apparent darkness is often an inversion effect, not an absence of energy.
  A useful analogy is photographic negatives: a bright source in a positive image becomes dark in the negative, even though the underlying information hasn't vanished. G-Ion dynamics can produce similar inversions. Energy is present and flowing, but its mode of interaction makes certain regions appear dark to our instruments.
- There is no truly empty space between particles.
  Even the gaps between quarks, leptons, and other subatomic entities are filled with G-Ion connectivity. The fabric is continuous, not granular.

Within this fabric, different classes of G-Ion play distinct roles:

- Active G-Ions
  These have polarity arrangements that allow perfectly balanced magnetic vertices. They can lock together into stable geometries.
- Non-active G-Ions
  These fail to achieve perfect polarity balance on their own and often function as connectors or spacers, helping maintain continuity across the fabric.

- Re-active Unstable G-Ions
  These include shapes such as tetrahedral G-Ions that can't settle into stable polarity. Their vertices constantly flip between positive and negative states. They can still form enclosed clusters, but only while perpetually shifting.

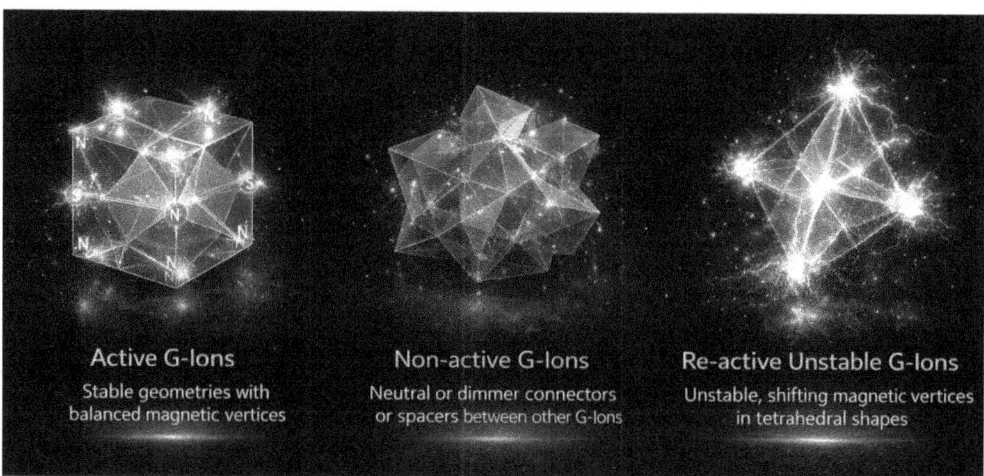

It's this last group, the Re-active Unstable G-Ions that provides the engine for perpetual motion. Their inability to stabilise drives constant interaction, friction, attraction, repulsion, and collision. The fabric never fully rests.

From this continuous motion, familiar phenomena emerge:
- the ageing and decay of matter,
- energy release through combustion and thermal processes,
- subatomic interactions and particle creation or annihilation,
- the immense power of supernovae and black-hole events,
- and the ongoing dynamics of gravitational fields.

Later sections, including the *Jelly Mould* discussion in **Gravity Matters**, explore how these G-Ion geometries generate gravitational wells and large-scale distortions. For now, the key takeaway is this:

G-Ions are not just another particle. They're the proposed geometry, energy, and connective fabric of the Omniverse itself, from which all structure and behaviour emerge.

REVIEW AND CONTINUATION OF THE G-ION THEORY

Theory progression from 25/10/1994 paper by C. Daniels

In the early stages of developing the G-Ion theory, two questions arose almost immediately.

First, the **shape problem**: what does the particle actually look like when it exists beyond the reach of quantum-scale observation?

Second, the **motive problem**: how does such a particle behave, move, and reconfigure itself under the influence of its own magnetic geometry and the surrounding G-Ion fabric?

Over time, those early uncertainties began to resolve. With continued exploration, a far more complete picture emerged, one that positions the G-Ion as a plausible foundational structure for all matter and antimatter. There's still work to be done, of course. As with many frontier theories, direct verification may depend on future technological advances. Even so, the internal logic, theoretical consistency, and alignment with observed anomalies place the theory on firm conceptual ground.

Later in this chapter, I examine String Theory in more detail. It's an ambitious and mathematically elegant framework that has pushed theoretical physics into genuinely new territory. Its limitation isn't a lack of mathematical sophistication, but a constrained conceptual frame. Repeatedly, modern physics attempts to describe an effectively infinite Omniverse through a narrow observational keyhole. No matter how advanced the mathematics, a framework bounded by finite assumptions can only ever capture part of the whole.

If the Omniverse is not a contained object. It has no boundary, no edge, no perimeter, no external reference point. It simply *is*. Eternal and endlessly extensible. Any theory that ignores this is structurally handicapped. Even our local cosmic universe is almost certainly just one bubble among countless others, nested within larger layers of structure. The conclusion is confronting but unavoidable: an infinite system cannot be fully described using a closed, finite rule set.

This is why theoretical models so often fracture at their limits. It's not that the mathematics fails, but that the context is too small.

Once Eternity is treated as a foundational condition of physics rather than a philosophical aside, the picture changes dramatically. Missing elements in String Theory stop looking like errors and start appearing as incomplete parameters. When the conceptual frame is widened to include infinite scale, infinite divisibility, and infinite dimensional layering, the G-Ion hypothesis fits naturally into place.

It's within this broader context that the next phase of the theory emerged.

The original 1994 paper proposed that if a particle existed below the quantum scale, one capable of stretching, contracting, and shifting energetic state on demand, it could function as the underlying geometry of all matter. That particle is now identified as the G-Ion. It also appears to align with the elusive particle implied in early investigations of **Transverse Magnetic Poling (TMP)**. In this revised framework, the terminology is unified: the particle is the G-Ion.

The theory also supports a significant extension: light speed is not the ultimate velocity limit of the Universe.
If G-Ions form the fabric of space, energy, matter, antimatter, dimensional states, and ethereal layers, then thought itself, as a property of that fabric, can navigate through it in a manner analogous to light. Possibly far faster.

Possibly without velocity at all. Within this framework, *Thought Waves* and *Thought Particles* become as relevant to physics as photons and electromagnetic radiation.

Revisiting the 1994 work through this expanded lens reveals a striking implication: neither the smallest nor the largest scale can ever be fully known, because both collapse into the reality of infinity.

There will always be a smaller "smallest" particle and a larger "largest" structure. This isn't conjecture. It's a direct consequence of an infinite Omniverse. And because the Omniverse is eternal, inwardly and outwardly, G-Ions remain both smaller and larger than anything we can measure, depending on state and function.

This distinction is particularly relevant for those working within String Theory. The theory is brilliant, but incomplete. It describes the behaviour of strings, not the architecture that gives rise to them.

G-Ions supply that architecture.

They provide geometry, dynamics, expansion and contraction mechanisms, magnetic poling, energy generation, and infinite scalability, all elements String Theory requires but cannot supply on its own. When integrated, the two frameworks produce a far more coherent picture of how the Universe is constructed, and they point toward tangible pathways forward in areas such as TMP, gravitational manipulation, advanced energy systems, and genuine mass-distance space travel.

In short, the G-Ion theory doesn't replace String Theory. It completes it. And in doing so, it opens the door to an entirely new frontier of physics.

A G-Ion doesn't move in the way classical or quantum particles do. It doesn't travel from point A to point B. Instead, it transfers itself by passing its complete informational blueprint, geometry, polarity configuration, energetic state, and dimensional signature, into the surrounding G-Ion fabric.

Each receiving G-Ion becomes an exact clone of the original at that moment, then immediately passes that blueprint onward, forming a continuous relay of perfect metamorphic replication.

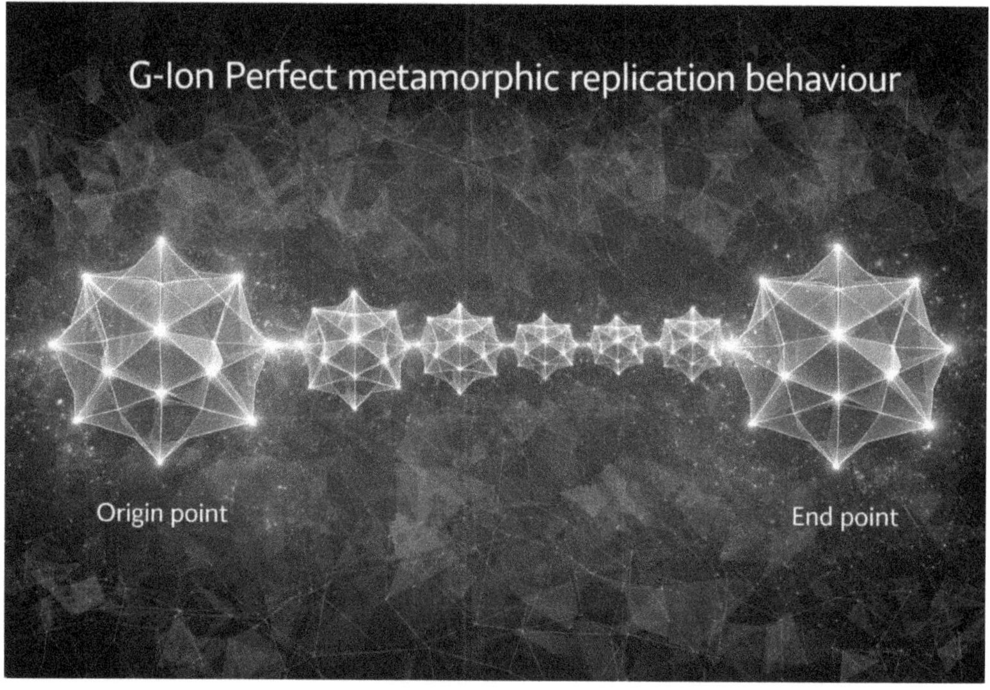

This imprint-transfer behaviour allows a G-Ion to *appear* to traverse any distance, even across unimaginable scales, without leaving its original position. The G-Ion at the destination is not the same particle, yet by every measurable criterion it *is* the same particle, because it embodies the exact state of the original at the moment of transfer. In practical terms, the starting G-Ion has occupied a new location without displacement.

This mechanism underpins the ability of matter within the G-Ion fabric to exhibit non-locality and apparent multi-position occupancy.

Why this process is never static

Because every G-Ion exists within a dense, interconnected ocean of other G-Ions, none can maintain a fixed shape or scale for more than an infinitesimal interval.

Each G-Ion is constantly negotiating magnetic residency, resisting displacement, and responding to the pressures of the surrounding fabric. The result is perpetual motion, friction, and transformation.

A G-Ion may shift from sub-quantum scales to astronomical dimensions and collapse back again within timeframes far below current measurement capability. This constant recalibration allows the G-Ion to remain locally coherent while functioning as both transmitter and receiver of imprint information.

Imprint transfer as distributed existence

Because the originating G-Ion immediately reshapes itself after transfer, there is never a permanently fixed copy.

The G-Ion at the destination is no longer identical to the source, even though both were identical at the moment of replication. In this way, G-Ions behave less like discrete particles and more like adaptive nodes within a fluid matrix.

This may offer one of the earliest physical clues toward unifying thought, information, and geometry at the deepest layer of reality. If the Omniverse operates through imprint transfer rather than particle displacement, then instantaneous communication, non-locality, and simultaneous presence cease to be paradoxes. They become natural consequences of the medium itself.

On geometric identity

G-Ions may adopt any geometry required by local conditions, but perfect symmetry is virtually impossible due to perpetual motion and polarity negotiation. A G-Ion may appear spherical at a distance, but closer inspection suggests a geodesic polyhedral structure, with faces continually shifting between triangular, hexagonal, and other polygonal forms.

This aligns with the central premise of the theory: nothing within an eternal, dynamic fabric remains static or perfectly symmetrical.

G-Ions, String Theory, Unification and Dark Energy

The G-Ion is more than just another proposed particle. In this theory, it's the missing piece that binds and unifies matter, so-called anti-matter, and the behaviours we associate with light, thought, dark energy, gravity, and the structure of the Omniverse itself.

The G-Ion framework aims to offer a single, coherent picture that can, at least in principle, address many of the long-standing open questions in physics. That includes the nature of light and thought, the possibility that all time exists as a single non-linear moment, the construction of our local cosmic universe, and the wider eternal Omniverse that contains it. It also reaches into planetary and cosmic chemistry, biology, material and elemental behaviour, and even the suggestion of an architectural responsiveness within the fabric of existence.

It's a large scope, deliberately so.

String Theory offers a useful point of comparison. Since early work in the 1920s and its later formal development, it has sought to unify fundamental forces by treating particles as vibrating strings.

Enormous intellectual effort has gone into that approach, and much of it is genuinely impressive. Yet even its strongest advocates acknowledge that there are still gaps and unresolved connections. In practice, String Theory often functions as a form of back-engineering: starting with observed behaviours and refining the mathematics until the model aligns.

The G-Ion theory proposes a different starting point.

Rather than stretching existing models toward unification, it assumes a deeper fabric that already contains what we recognise as mass, matter, anti-matter, space, apparent empty space, energy, the Higgs field, quarks, gluons, protons, neutrons, electrons, strings, and vibrational resonance. In this view, these are not separate layers to be bolted together, but downstream behaviours emerging from G-Ions and their interactions. Within this framework, G-Ions occupy all three primary layers of matter described earlier:

- **Material** – solid, liquid and gaseous matter
- **Prothereal** – non-solid structural fields and what we interpret as "empty" space
- **Ethereal** – subtle or non-conforming matter, often misidentified as "nothingness" or dark phenomena

They also permeate all six primary dimensional realms: Time, Space, Distance, Thought, Existence and Eternity.

Because of this, a G-Ion isn't pinned to a single size or state. It can expand or contract effectively without limit, and transition between extreme cold and extreme heat almost instantly. This allows G-Ions to span super-massive distances and volumes, or collapse into unimaginably small scales, at rates far beyond our current observational reach.

It's helpful to think of a G-Ion as a frame rather than a solid object.

That frame can adopt almost any geometric configuration, stretching, skewing and reforming as conditions demand. Inside the frame, depending on its role, exists Prothereal or Ethereal matter.

When G-Ions pack tightly into highly compressed clusters, the result is what we experience as solid Material matter. It feels stable, yet we know even the most rigid forms can release vast energy under the right conditions. That instability is intrinsic to the G-Ion fabric.

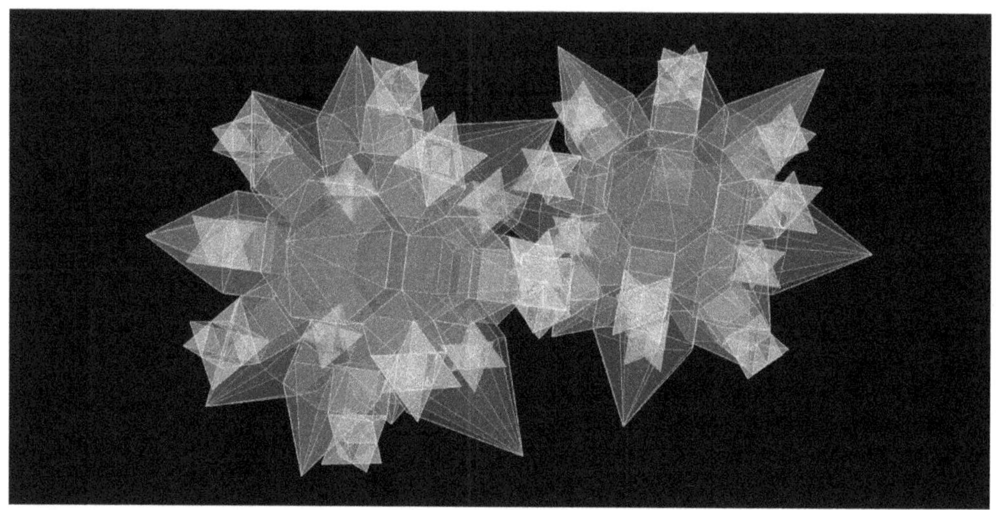

Artistic impression of G-Ion clusters

From this perspective, what we label dark energy or dark matter largely reflects Ethereal G-Ion states and the way G-Ions store, release and redirect energy within regions we currently treat as empty. The invisible scaffolding of the universe is doing far more work than we usually account for.

This is where String Theory and G-Ion theory naturally overlap.

Each line segment, or edge, of a G-Ion frame functions as a string. These strings have indeterminate length, carry vibrational energy, and meet at vertices that hold magnetic polarity.

The faces of a G-Ion may be square, triangular, hexagonal, or more complex geometries, each with its own arrangement of positive and negative poles at the corners. The result is a constant negotiation for magnetic residency, both within individual G-Ions and across the surrounding fabric.

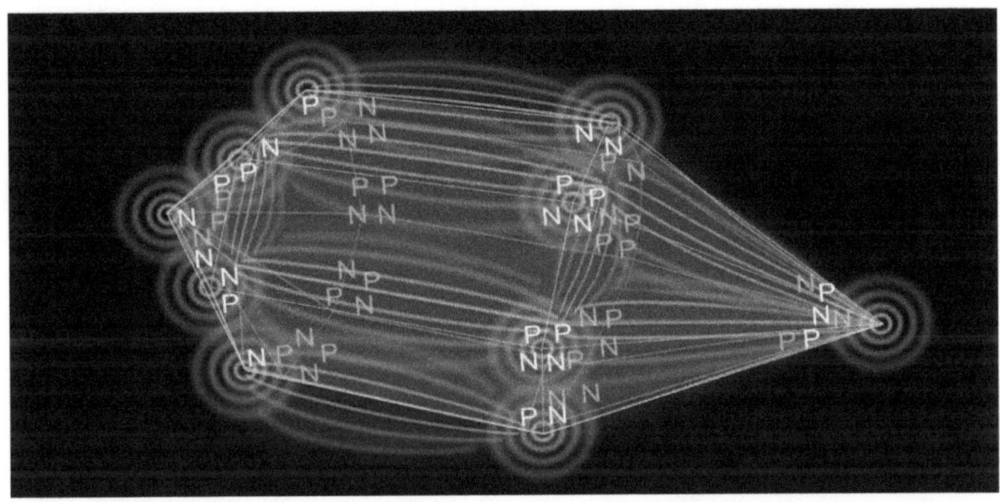

G-Ion cell with vibrating strings and vertices producing electromagnetic friction caused by the Positive and Negative residency battle for polarity dominance.

Where conventional notation uses terms such as *shm* (simple harmonic motion) or *sho* (simple harmonic oscillator), G-Ion modelling places greater emphasis on internal vibrational oscillation within geometric frames. For that reason, I've used **ʋo** or **ʋoʋ** as shorthand for vibrational oscillatory behaviour. It's not standard physics notation, but it provides a practical way to track internal dynamics within G-Ion formulations.

In short, if G-Ions exist as described, they offer:

- A physical basis for string-like behaviour,
- a way to unify dark energy, visible matter, and the space between space,

- and, a candidate mechanism for unification operating from the smallest scales to the Omniverse itself.

Whether the theory withstands experimental scrutiny is a question for the future. For now, its value lies in offering a coherent framework that can, at least in principle, bring string theory, unification concepts, and dark energy into a single, connected conversation.

COSMIC DISTORTION AND THE UNIVERSAL CHINESE WHISPER

When we push deeper into subatomic and quantum territory, we repeatedly encounter behaviours that don't sit comfortably within our everyday models of space, time and gravity. From the G-Ion perspective, this isn't a flaw in nature, it's a signal that we're only seeing surface effects cast by a deeper fabric beneath.

In this framework, G-Ions sit below the familiar quantum zoo. They form the base texture from which quarks, gluons, photons, bosons and gravitons emerge. Because G-Ions define the geometry and energetic behaviour of the fabric itself, their dynamics force us to rethink some of our most basic assumptions about space, structure and interaction. Their energy limit is unknown, not because it's infinite by declaration, but because their behaviour isn't constrained in the ways we currently know how to measure.

One of the primary motivations for laying this out so carefully is gravity.

We still don't have a fully unified, experimentally complete description of gravity that integrates cleanly with quantum mechanics. General relativity gives us an elegant and extraordinarily successful description of curved space-time at large scales, and within that domain it works brilliantly. But as soon as we try to extend that picture into an omni-directional, multi-layered Omniverse, its limitations begin to appear.

If G-Ions form the actual fabric, then gravity isn't a separate field acting on that fabric. Gravity is what the fabric does when it's distorted, stretched, compressed and driven into complex motion. That distinction matters, because it changes how we interpret nearly everything we observe.

Take light as an example. We often model photons as following neat geodesic paths through curved space. But if light is both attracted to and repelled by electromagnetic structures within the G-Ion fabric, then what reaches our detectors may be a heavily altered version of the original signal. By the time that light arrives, it may have been bent, delayed, magnified, diminished or filtered countless times as it passed through gravitational wells, energetic currents and structural distortions.

If that's even partly true, then a significant portion of our cosmic picture is based on information that's already been transformed by a medium we don't yet fully understand.

That doesn't mean existing cosmology is wrong, dishonest or useless. It means our calculations are only ever as good as the signals we receive, and those signals may already be distorted before we measure them. In effect, we may be reconstructing the universe from a translated version of events, not the original.

A simple analogy helps here: Imagine the cosmos as a giant game of Chinese whispers. A message is passed from one participant to the next around an enormous circle. By the time it reaches you, the words have shifted. In much the same way, a photon that left its source billions of years ago has travelled through a constantly changing maze of G-Ion distortions, gravitational structures, electromagnetic interactions and space-time ripples. What finally arrives at our instruments may carry a faithful outline of the story, but not the exact phrasing of the original.

This doesn't mean we throw out cosmology and start again. It does suggest we should be cautious about treating current models as final. We need better ways to cross-check incoming data, and new methods for inferring what original signals likely looked like before they were reshaped by the medium they travelled through.

Viewed this way, many inconsistencies and oddities in cosmic mathematics and observation become easier to understand. The problem may not be the mathematics itself, but the fact that we're feeding it incomplete or distorted input.

In that context, standard ideas such as the Big Bang begin to look less like ultimate answers and more like provisional working models. They may describe certain observed patterns accurately while still missing deeper mechanisms operating within the G-Ion fabric that produced those patterns in the first place. The intention here isn't to dismiss mainstream cosmology, but to acknowledge that our view is partial, filtered and sometimes misleading, and that the G-Ion perspective may help explain why.

From a practical standpoint, this line of thinking also opens new possibilities. If G-Ions are the true carriers of energy and structure, and if their motion is genuinely perpetual across vast timeframes, then technologies that interact directly with that behaviour could, in principle, produce extremely long-lived energy systems.

Concepts such as advanced photonic energy extraction or transverse magnetic poling may eventually emerge not as isolated curiosities, but as applied expressions of this deeper fabric.

With that in mind, gravity no longer needs to be treated as a mysterious force layered on top of reality. It becomes a natural consequence of how the G-Ion fabric moves, compresses and negotiates itself.

That shift in perspective is where we turn next, by looking at gravity not as something that acts on matter, but as something matter does.

Gravity 'Matters'

The G-Ion Jelly Mould Theory

If the G-Ion theory is correct, then gravity is not merely a force acting at a distance. It's a form of energy matter, an intrinsic behaviour of the underlying fabric of existence itself.

One of the major limitations in our attempts to understand, and eventually manage gravity, is that we've treated it as intangible, something that acts *through* space rather than *as part of it*.

Gravity appears invisible only because we observe its effects, not its substance. If, instead, we view gravity as a continuous behavioural state within the G-Ion fabric, its behaviour becomes far easier to interpret. What we describe as variations in gravitational strength are not arbitrary. They're direct responses to the size, mass, volume, density and autonomous energy of the bodies embedded within that fabric.

Gravity is not generated by planets and stars. It already exists as a property of the G-Ion fabric. Planets and stars do not create gravity; they displace and organise it. If gravity did not pre-exist as a behavioural property of the fabric, stars could not condense from hydrogen clouds, and galaxies could not cohere in the first place.

The G-Ion fabric does not arise from gravity. Gravity arises from the G-Ion fabric. Even in the absence of gravitational behaviour, the fabric would remain. The reverse is not true. Without the fabric, there can be no gravity, no matter, and no physical effects whatsoever.

Gravity as a Substance of Energy Matter – Why Bodies Move Toward Each Other

In any gravitational environment, the forces we observe depend on the nature of the bodies occupying that environment.

A small planet produces one pattern of compression and displacement. A large star produces another. The gravitational energy matter attempts to reclaim the space it occupied before the physical body arrived, and this produces behaviours of contraction, stretching and squeeze around that body.

This explains why larger and denser objects generate stronger gravitational effects, and why gravitational influence extends outward to a point of equilibrium where compression can stabilise.

People often say, "There's no gravity in space." The reality is the opposite. Space itself participates in gravitational behaviour. What astronauts experience is not an absence of gravity, but the absence of a steep gravitational gradient.

An object within a gravitational environment doesn't experience gravity uniformly. Compression is stronger on the side closer to a massive body and weaker on the side facing away. These stresses do not cancel. The result is a net acceleration toward regions of higher compression.

Motion arises not because gravity pulls, but because the surrounding energy matter is unevenly distributed across the object's volume.

Smaller bodies respond more strongly to these gradients, while larger bodies respond more weakly due to greater inertia. Both bodies, however, contribute to and respond to the shared compression pattern.
What we call gravitational attraction is simply the natural motion of matter embedded within an unevenly compressed medium.

The Jelly Mould Analogy

Imagine gravity as a vast, spherical jelly mould made of energy matter. At first, there are no planets or stars inside it, only the jelly itself, evenly distributed and in equilibrium.

Now imagine inflating a ball at the centre, representing a star or planet. As the ball expands, it displaces the jelly outward. Near the ball, the jelly becomes tightly compressed. Further away, compression gradually relaxes until the jelly resumes its original density.

Illustration: Although shown here in cross-section for clarity, gravitational compression in the G-Ion ocean is omnidirectional and equal in all directions.

If additional balls are introduced, representing stars, planets, moons or asteroids, each displaces the jelly in proportion to its mass, volume and density. Planets and stars do not generate gravity. They distort a gravitational medium that already exists.

The gravitational field is simply the jelly being displaced, stretched and compressed as it attempts to return to equilibrium.

*Simple illustration showing the jelly-mould gravitational field
interactions around the Sun star and Earth*

What Gravitational Pull Really Is

In this framework, gravitational pull is shorthand for a deeper process.
Motion arises from a compression gradient that produces unequal stress
across an object, resulting in acceleration toward regions of greater
compression.

For readers familiar with conventional physics, this maps directly onto
gravity expressed as motion driven by a potential gradient:

$$F = -m\nabla\Phi$$

Here, Φ represents gravitational potential. In jelly terms, Φ corresponds
to local compression, and $\nabla\Phi$ describes how rapidly that compression
changes with position. Objects accelerate toward increasing compression.

The Rules of the Jelly

To make multi-body gravity inevitable rather than mysterious, the jelly follows a small set of simple rules:

- Space (the G-Ion fabric) behaves as an elastic medium seeking equilibrium.
- Massive bodies displace this medium, creating local compression.
- Gravitational strength corresponds to how steeply compression changes with distance.
- Objects accelerate because compression is unequal across their volume.
- Each body produces its own squeeze pattern; nearby patterns superimpose.
- Motion follows the gradient of the combined compression field.
- Tidal effects arise wherever that gradient varies across an object's size.

Gravity becomes the natural behaviour of matter embedded in a medium that is continually attempting, and failing, to return to balance.

Tides Explained Naturally

Because a planet's compression gradient varies across the diameter of a moon, the near side experiences stronger compression than the far side. The result is stretching along the connecting axis.

The same effect occurs on planets with oceans. Water on the side facing a moon sits deeper in the compression gradient than water near the planet's centre, while water on the far side sits in a region of comparatively lower compression. The result is two tidal bulges formed by unequal compression across the planet's body.

Tides are not caused by a direct pull on water. They're caused by differential compression across the planet's body.

Anomalies: Gas Giants and Compact Stars

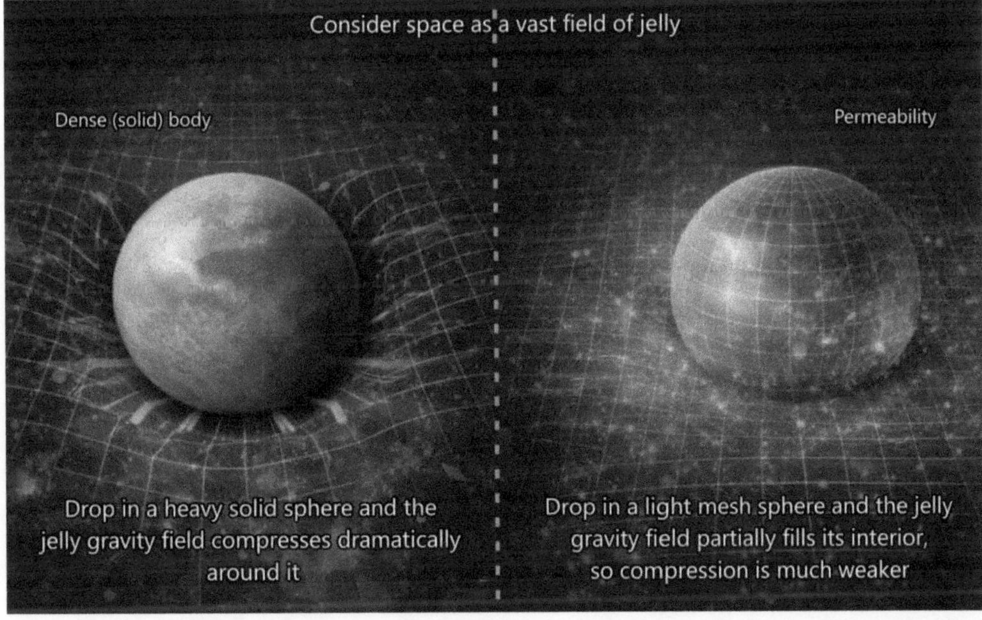

Replace a solid ball with a wire-mesh ball, representing a gas giant. Because internal density is lower, the jelly can partially infiltrate the interior, producing a different compression profile.

Stars behave differently again. A star produces autonomous energy, heat and radiation. This output prevents the surrounding gravitational medium from settling, creating continual collapse-and-restoration cycles that amplify gravitational effects.

When a massive star collapses, compression reaches an extreme. The surrounding medium rushes inward. Under sufficient pressure, this produces black holes.

Note: Although some illustrations may resemble curved-space diagrams, they are intended to represent pressure, density and behavioural gradients within a continuous G-Ion medium, not geometric deformation of empty space.

Black Holes: G-Ion Collapse, Not Infinite Density

In the G-Ion framework, black holes do not require infinite density. They represent total collapse of gravitational energy matter to a forced equilibrium state.

That equilibrium need not be permanent. A perturbation could, in principle, trigger catastrophic release events. The model does not require a single Big Bang. It allows Big-Bang-like events as natural outcomes of extreme compression and release.

Returning to the G-Ion Perspective

If gravity is energy matter, it fits naturally into the wider G-Ion framework. If the G-Ion ocean is the seamless substrate from which matter, energy and interaction arise, then gravity is a behavioural expression of that fabric as it stretches, compresses, collides and reorganises in an effort to maintain equilibrium.

Considering gravity this way opens an entirely new field of possibility. Instead of attempting to manipulate gravity as an external force, we can begin to study its internal mechanics. Once the behaviour of the fabric is understood, gravitational dynamics become something we can, in principle, interact with and influence.

From Structure to Motion

So far, we've focused on what gravity *is*. But structure alone doesn't explain motion.

Planets orbit. Moons migrate. Stars wobble. Galaxies rotate in ways that resist simple explanation. Gravitational systems evolve, resonate and destabilise.
If gravity is the behavioural response of the G-Ion fabric, then gravitational dynamics describe how that response unfolds over time.

Einstein showed us that freely falling objects feel no force at all. In the G-Ion framework, this insight deepens: objects move not because they're pulled, but because the fabric around them is reorganising.

To understand gravity fully, we must stop thinking in static terms and begin thinking in behaviour, flow and response. With that shift, gravity stops being something that merely exists and becomes an active, state-dependent behaviour of the fabric, responding to conditions..

This is where gravitational dynamics begin.

GRAVITATIONAL DYNAMICS

HOW G-ION BEHAVIOUR CREATES THE PHENOMENA WE CALL GRAVITY

If gravity is not an invisible force pulling objects together, but the behaviour of the G-Ion fabric responding to mass, energy, motion and pressure, then the next step is to understand what that behaviour looks like in motion. This is where gravitational dynamics enter the picture.

We're accustomed to gravity being described in terms of curved space, bent geometry or falling reference frames. Those descriptions, developed through Einstein's general relativity, remain extraordinarily successful within their domain. They describe what we observe. They do not, however, describe the underlying mechanism.

Under the G-Ion framework, gravity is not a consequence of curved space-time. Curved space-time is a consequence of G-Ion behaviour.

Rather than imagining empty space being bent like a rubber sheet, imagine a densely woven, perpetually active G-Ion ocean.

Every object, from a dust mote to a star, interacts with this fabric by displacing it, compressing it, or setting it into motion. The gravitational field we observe is simply the visible outcome of how the G-Ion fabric reorganises itself in response to that disturbance.

ALL SPACE IS NOT "SPACE"

Space, in the G-Ion framework, is not an empty void. It's a continuous, active G-Ion medium, characterised by variations in density, pressure and dynamic behaviour. No geometric curvature is implied here. All apparent structure arises from internal reorganisation of the fabric itself.

The G-Ions' constant stretching, contracting, vibrating and reconfiguring create an active medium with viscosity, density and continuous energy flow, even if those properties sit far beyond the reach of our current instruments.

Within this view, gravity is not simply a force acting on objects. It's the behavioural response of the G-Ion fabric as it adapts to mass, density, volume and autonomous energy.

Although matter and gravity appear distinct to us, within the G-Ion framework they're two expressions of the same continuous medium. Matter represents a constrained, coherent configuration of G-Ions. Gravity represents the dynamic response of nearby G-Ions displaced from their equilibrium state by that constraint. The fabric is continuous. Only its behaviour changes.

Traditional physics often describes gravity as the curvature of space-time. The G-Ion framework reframes this in mechanical terms.

Gravity arises from how the G-Ion fabric adapts when matter occupies configurations that disrupt its internal balance. Mass does not pull space toward itself. Instead, the G-Ion fabric compresses, redistributes and rebalances around structured regions. The degree of compression depends on mass, volume, density and internal energy.

Objects moving within these gradients experience what we label gravitational acceleration. Importantly, the G-Ion fabric does not flow around an object. The object itself is a structured configuration of that same fabric.

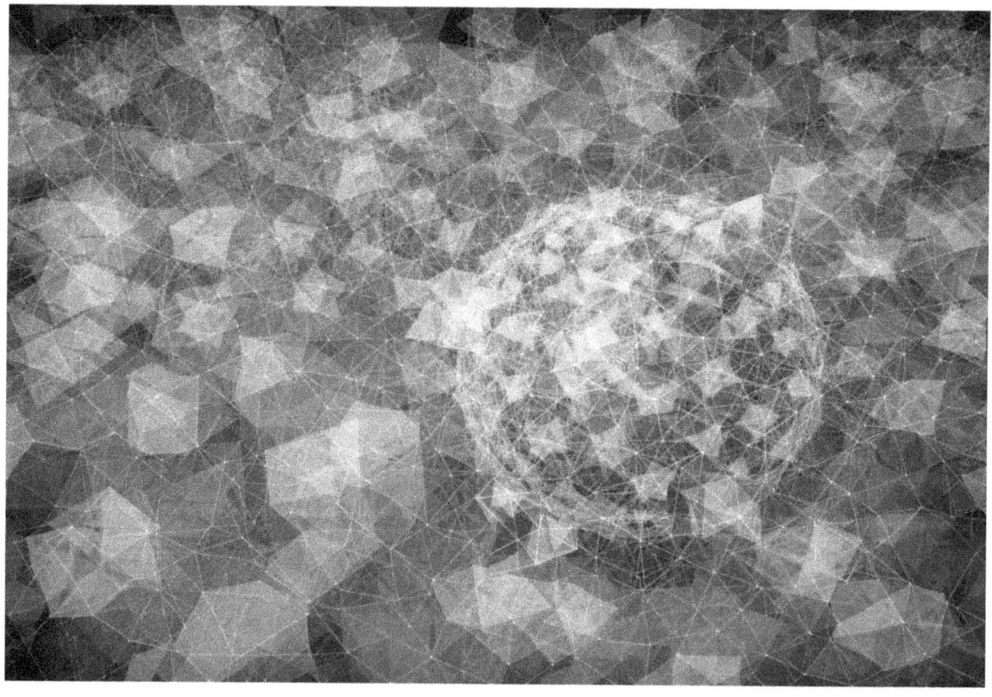

Solid Body formation as a Structural G-Ion configuration

In the illustration, a dense object appears as a region of tightly constrained G-Ion cohesion. The surrounding fabric is not displaced externally, but internally reorganised, producing gradients of pressure and stress that propagate outward through the continuous medium. What we experience as a gravitational field is the behavioural response of surrounding G-Ion dynamics as they are displaced from their preferred equilibrium by that configuration.

A useful analogy is placing a dense object into an ocean. The water compresses, redistributes and continually attempts to stabilise. The denser the object, the greater the displacement. If the object vibrates, rotates or heats the surrounding water, the behaviour changes again. Gravity behaves in a similar way, except instead of water, we are dealing with a reactive, polarised, hyper-elastic G-Ion medium.

Under this model:
- Massive bodies produce stronger compression and reorganisation of the G-Ion fabric.
- These changes generate pressure and density gradients we interpret as gravitational fields.
- Smaller bodies move along paths of lower resistance within those gradients.

Notably, this aligns with Einstein's insight that a freely falling object feels no force at all. Under the G-Ion model, nothing is pulling on it. It simply follows the path of least resistance through the G-Ion field.

This is not a contradiction of mainstream physics. It is a different descriptive language for the same class of observed behaviour.

DENSITY, COHESION AND THE G-ION LATTICE

The density of a physical object arises from the degree of G-Ion cohesion within it.

Stronger cohesion produces denser materials such as rock, metal or stellar cores. Weaker cohesion produces lower-density states such as gas giants, nebulae or diffuse plasma. This variation in cohesion directly influences how the surrounding G-Ion fabric reorganises.

Partial Displacement in Low-Density Bodies
(Spherical geometry is for illustration purposes only)

A dense body restricts G-Ion motion and produces steeper pressure gradients. A less dense body allows partial permeation, softening those gradients.

Density, then, is not merely an emergent property of matter. It's a direct expression of how G-Ions are grouped, constrained and permitted to move.

G-Ion Restorative Stress

Within the G-Ion framework, the jelly is not passive. It is vibrating, reactive, magnetically polarised and perpetually reshaping.

When structured regions of the G-Ion fabric emerge as matter, they introduce localised stress, polarity shifts and density gradients that displace equilibrium in the surrounding medium. This displacement is not external. It's intrinsic to the fabric itself.

The resulting **G-Ion Restorative Stress** drives the medium toward reconfiguration, producing pressure flows that manifest as gravitational behaviour.

Gravity, in this sense, is not a pull, but the visible outcome of the G-Ion fabric continually attempting to restore balance around constrained configurations.

Orbital Motion as Dynamic Equilibrium

Within this framework, orbital motion emerges naturally.

An orbit represents a dynamic equilibrium between a body's tangential motion and the surrounding G-Ion restorative stress gradient.
Objects do not remain in orbit because they're being pulled inward, but because their motion continually traces paths of stable resistance within the G-Ion field.

Circular orbits correspond to near-constant restorative stress contours. Elliptical orbits reflect oscillation through regions of higher and lower stress density. Orbital eccentricity, in this view, is a measure of asymmetry in the surrounding G-Ion stress landscape.

This reframing preserves all observed orbital behaviour while removing the need for gravity to act as a distant force or purely geometric abstraction.

DYNAMIC, NOT STATIC: WHY GRAVITY IS ALWAYS CHANGING

In classical models, gravity is often treated as smooth and static. In the G-Ion framework, variability is not an anomaly. It's expected.

Any change in density, temperature, motion or electromagnetic output alters how the surrounding G-Ion fabric must reorganise. A star's pulsation, a planet's molten core, or tidal interactions between bodies all generate shifting restorative stress patterns.

These shifts manifest observationally as orbital anomalies, gravitational waves, frame-dragging effects, lensing distortions, unexpected galactic rotation profiles, and cluster-scale behaviours currently attributed to dark matter.

From this perspective, dark matter may not be hidden mass at all, but unmeasured variations in G-Ion density, cohesion and restorative stress across large cosmic regions.

THE ROLE OF HEAT, MOTION AND CHARGE

A hot, turbulent object such as a star continuously agitates the G-Ion fabric, preventing full stabilisation and producing dynamic stress fields with persistent ripples and resonances.

In conventional terms, this resembles:
- magnetohydrodynamic turbulence
- plasma field distortion
- flux rope formation
- variable pressure zones in solar wind and magnetospheres

From a G-Ion perspective, these are simply observable consequences of the fabric attempting to restore equilibrium while internal conditions continually disrupt it.

Colder, more stable bodies generate smoother, more predictable stress fields. This helps explain why neutron stars, black holes, gas giants and solar-mass stars exhibit vastly different gravitational behaviours, despite all being described as "massive".

It is not just the amount of mass that matters, but how the G-Ion fabric reorganises in response to density, heat, internal motion, electromagnetic resonance and rotational velocity. These factors collectively shape each gravitational environment.

A UNIVERSE THAT IS NOT A VOID, BUT A SINGULAR FIELD

In this view, the universe is not a collection of objects floating in an empty container. It's a continuous, cohesive field composed of:

- dense regions such as stars, planets and cores
- sparse regions such as nebulae and apparent voids
- and the all-pervading G-Ion ocean that binds them

Much like a human body is a single organism composed of countless autonomous cells, the cosmos is a singular system composed of countless G-Ion configurations, each interacting locally while contributing to the behaviour of the whole.

TOWARD GRAVITATIONAL MANIPULATION, AND BACK TO THOUGHT

If gravitational behaviour arises from G-Ion dynamics, then interacting with gravity means interacting with the G-Ion fabric itself.

This doesn't imply fantasy technology. It mirrors how understanding electromagnetism unlocked motors, radios and modern electronics. Once a phenomenon is recognised as a structured, interactive medium rather than an abstract effect, entirely new methods of engagement become possible.

If gravity behaves as a dynamic field of energy matter rather than a fixed geometric abstraction, then new approaches to propulsion, energy transfer and inertial control become at least thinkable.

And this brings us full circle.

If gravity is the behaviour of the G-Ion fabric, and if thought also interacts with that same fabric, then gravitational dynamics and cognitive dynamics may not be separate domains at all. They may be two expressions of the same underlying medium.

In that sense, gravitational dynamics is not just a physical theory. It's a doorway. A doorway leading toward the next stage of this journey: how interaction with gravity, space, time and matter may eventually occur through the only mechanism subtle enough to operate within the G-Ion domain.

Thought.

For readers who wish to see how this perspective maps onto familiar gravitational and orbital language, a short, math-light appendix is included at the end of the book.

Our Mind - the Only True Instrument of the Omniverse

If the G-Ion fabric is truly the foundational mechanism beneath all matter, force, structure and continuity across the Omniverse, then we face an unavoidable implication. G-Ions may be fundamentally unreachable by any mechanical device we can ever build. No machine, probe, accelerator, sensor or instrument, no matter how advanced, will ever be small enough, fast enough or subtle enough to directly observe, penetrate, interact with or manage the G-Ion ocean. Every technological tool is made of G-Ions, which means every tool exists inside the very system it is attempting to measure.

This is not a technological failure. It's an architectural boundary built into existence itself.

But this doesn't mean the G-Ion fabric is forever beyond our reach. It simply means we've been looking in the wrong direction. The right instrument is not mechanical. It's cognitive. The only structure subtle enough to operate at the G-Ion scale is thought, guided by the human mind.

When I refer to the mind, I'm not talking about the biological brain as a computational machine of neurons, chemistry and synaptic firing. I'm talking about the extended system the brain enables. The mind is a vibrational pattern, a transferable structure, a navigational state capable of imprinting, influencing, receiving and responding. It operates across both the internal network and the wider informational field. Thought is the information exchanged through that network. Despite appearances, thought is no less tangible than the electrical activity moving through the brain.

This invites a shift in how intelligence, creativity and imagination are understood.

If thought is a physical mechanism operating within the G-Ion fabric, then it's not confined by mass, inertia or material boundaries. It already behaves as a native process within the Omniversal architecture. Thought interacts with the G-Ion ocean because it arises from the same underlying substance. It operates using the same internal language of reality.

If this is true, then the most powerful scientific instrument we will ever develop may not be a device. It may be ourselves.

Our minds may already be the most advanced instrument for exploring the deeper structure of existence. The real challenge is learning how to use the entire operating system.

At present, we resemble someone who's just powered on a new machine. We understand the basics, but it takes time, discipline and guided development to unlock its full capabilities.

For scientists, this reframes the frontier. Instead of endlessly building smaller probes to look deeper, the question becomes: what happens when the observer is already made of the thing being observed?

For philosophers and futurists, it suggests we do not merely inhabit the Omniverse. We participate in its structure.

For artists and creatives, it proposes that imagination is not symbolic alone. It may be structural information interacting with the cosmic framework.

For leaders and policymakers, it hints that the next true revolution will not be mechanical, digital or even quantum. It will be cognitive. It will arise from understanding and refining the one physical process that already operates at G-Ion scale.

And for the general reader, it offers something both simple and profound.

The mind may not just exist within the Universe. It may be capable of navigating it.

The next great leap in human progress may not come from building machines that probe deeper into matter, but from refining the instrument of the mind itself. When humanity finally understands thought as a physical mechanism rather than a purely psychological experience, we may gain access to a domain of interaction no telescope, collider or quantum sensor could ever reach.

This is not mysticism or metaphor. It's the logical progression of a species beginning to understand its own architecture. Thought, guided by the mind, may be the only mechanism subtle enough to operate directly within the G-Ion ocean.

Across the long arc of human evolution, we may discover that the ultimate interface with the Omniverse is not technological, but cognitive. We may never touch the G-Ion fabric with our hands, but one day we may learn how to shape it with intent.

And if this frontier is pursued with curiosity rather than fear, and discipline rather than superstition, interaction with the G-Ion fabric may one day become not speculative fiction, but a refined branch of physics. Not because we built the perfect tool, but because we became it.

FROM COGNITIVE INTERFACE TO ENGINEERED INTERFACE

If the mind is the only instrument subtle enough to interact directly with the G-Ion ocean, then a natural question follows: what role does technology still play?

Even if no device can penetrate or manipulate the G-Ion fabric itself, that doesn't mean engineering is sidelined.

Human progress has never depended on accessing the deepest layer of reality directly, only on learning how to shape the layers that emerge from it.

Electricity, magnetism, radio, semiconductor physics, fusion energy and quantum devices all operate by engaging with behaviours that arise from deeper structures, not by touching those structures themselves.

TMP-GEM, Transverse Magnetic Poling Gravitational Energy Management, belongs to this category.
It doesn't claim to manipulate the G-Ion fabric directly. No machine can. Instead, it attempts something far more grounded and defensible. It engineers macroscopic systems whose internal magnetic geometry reproduces the same behavioural conditions the G-Ion framework suggests underpin gravitational and energetic phenomena.

Rather than reaching downward into the sub-quantum ocean, TMP-GEM works upward from engineered electromagnetic fields, using controlled polarity asymmetry, magnetic vertex instability, geometric field compression and induced aposattraction behaviour to create conditions under which energy release, gravitational displacement and oscillatory resonance naturally emerge.

Where the previous section explored the cognitive interface, the possibility that thought may one day serve as a direct instrument of G-Ion interaction, TMP-GEM explores the engineering interface. It asks how geometry, magnetism and field dynamics can be used to harness the observable consequences of G-Ion behaviour without ever touching the G-Ions themselves.

This is the technological counterpart to the philosophical and cognitive ideas already explored.

A device cannot think, and therefore cannot participate in the G-Ion medium the way a mind potentially can. But a device can create instability, resonance and structured imbalance that mirror the dynamics shaping energy, inertia and gravity at deeper levels.

TMP-GEM is not a bridge into the Omniverse. It's an attempt to build a machine that behaves in harmony with it.

And like all meaningful engineering breakthroughs, it begins with a simple but radical question:

If gravity is the behaviour of a field rather than a force, what happens when we deliberately recreate that behaviour inside a machine?

That question led to the early sketches and schematic developments of Transverse Magnetic Poling, and to the system explored in the next chapter.

TMP-GEM

FROM SUB-QUANTUM PRINCIPLES TO PRACTICAL ENGINEERING

TMP-GEM (Transverse Magnetic Poling – Gravitational Energy Management) sits at the intersection of two domains: the invisible physics that shape reality, and the tangible engineering that allows us to interact with it.

Where the preceding chapters explored the G-Ion fabric as the foundational substrate of energy, matter and gravitational behaviour, TMP-GEM asks a more applied question: If we cannot yet manipulate the G-Ion ocean directly, can we build a physical device that recreates the same conditions under which G-Ion behaviour naturally emerges? This distinction matters.

TMP-GEM is not an attempt to manufacture the G-Ion fabric itself. Rather, it proposes that any engineered system capable of reproducing asymmetric electromagnetic pressure, polarity instability, vertex-level conflict and transverse field compression may naturally trigger the same class of behaviours the G-Ion ocean exhibits at the sub-quantum scale. TMP-GEM is therefore best understood as an engineering analogue.

It seeks to reproduce in matter what nature expresses in the G-Ion field:
- Continuous magnetic repolarisation.
- Oscillating pressure states.
- Pulse-wave field emissions.
- Gravitational inversion as a secondary effect.
- Sustained energy release as a by-product of continuous instability.

These behaviours are not speculative. Each has precedent within established physics, including:

- Ferromagnetic and multiferroic domain flipping,
- magnetostriction,
- piezoelectric stress–field coupling,
- plasma field resonance,
- and, non-linear electromagnetic instabilities.

TMP-GEM arranges these known principles into a geometry specifically designed to enforce a perpetual transverse poling cycle. In practical terms, the device forces magnetic domains into a state where:

1. They cannot settle into equilibrium.
2. They must continuously attempt to repolarise.
3. Every repolarisation produces a measurable energy pulse
4. The surrounding field undergoes detectable gravitational distortion.

From this point forward, TMP-GEM is treated not as speculative physics, but as a structured engineering proposal:

- What is TMP-GEM?
- How does it work?
- Why should it work?
- How do we build it?
- What materials are needed?
- What equations describe its behaviour?
- What is the path to a first functioning prototype?

TMP-GEM is where theoretical physics touches the workshop floor.

WHAT TMP-GEM IS

TMP-GEM (Transverse Magnetic Poling – Gravitational Energy Management) is a proposed engineering method for producing controlled energy release and localised gravitational manipulation by forcing magnetic fields into a deliberately engineered state of instability.

At its core, TMP-GEM rests on a single organising idea:

If gravity, inertia and energy are emergent behaviours of G-Ion reorganisation, then a physical system that recreates the same stress conditions and field instabilities may reproduce those behaviours at a macroscopic scale.

TMP-GEM does not claim to recreate or manufacture the G-Ion fabric. No machine can. What it aims to do instead is replicate the conditions under which G-Ion-driven behaviours naturally arise.

To do this, the concept employs:

- Electromagnetic asymmetry, using fields that are prevented from settling into equilibrium.
- Transverse magnetic poling, forcing opposing field geometries to intersect.
- Geometric cell structures, based on equilateral triangular grids that promote predictable instability.
- Aposattraction, engineered simultaneous attraction and repulsion.
- Controlled vertex polarity conflict, producing twelve-point instability within each triangular cell.
- High-strength, low-interference materials, capable of maintaining alignment under sustained stress.

TMP-GEM functions as an engineered **field stressor**.

The system consists of two hemispherical ERB housings (Electromagnetic Resonance Bells), each containing a precisely arranged lattice of magnetic cells. These housings are brought into transverse alignment such that their internal fields cannot resolve into a stable configuration.

When activated, the system enters a state characterised by:

- Perpetual magnetic repolarisation cycles, where positive and negative domains attempt to dominate but never succeed.
- Continuous vibrational friction, as magnetic vertices oscillate in search of polarity residency.
- Pulse-wave magnetic emission, generated by repeated failed equilibrium attempts.
- Local gravitational inversion, arising from field compression and release within the lattice.
- Usable clean energy, released as a by-product of continuous transverse repolarisation.

ERB Housing - Electromagnetic Resonance Bell

An ERB is an engineered hemispherical or shallow dome-capped cylindrical housing constructed from extremely high-strength, low-interference materials. Its function is to contain, stabilise and support the TMP-GEM cell lattice while withstanding the stresses imposed by transverse magnetic poling.

The ERB maintains precise geometric alignment, suppresses destructive field leakage, and preserves structural integrity as the internal magnetic fields undergo continuous repolarisation and vibrational friction cycles.

A SIMPLE EXPLANATION OF TMP-GEM

Imagine trying to push two strong magnets together, north-to-north or south-to-south. They fight. They vibrate. They resist. They snap sideways.

Now imagine thousands of micro-magnets arranged with geometric precision inside two hemispherical bowls, the ERB housings.

Then imagine forcing those bowls together with perfect alignment, so every one of those micro-magnets is trapped between wanting to align and wanting to repel.

That trapped instability produces energy. And when it's sustained, it produces a continuous cycle.

TMP-GEM is an attempt to do this in three-dimensional symmetry, using electromagnetic grids instead of permanent magnets. This allows controlled instability, tunable field strength, pulse-wave generation, and measurable gravitational side-effects.

In essence, TMP-GEM is a machine that uses the geometry of conflict to create usable energy and gravitational effects.

The scientific framing

While the terminology differs from mainstream physics, the underlying behaviours TMP-GEM relies on are not exotic:
- Field superposition.
- Opposing magnetic polarity interactions.
- Non-linear field resonance.
- Magnetic hysteresis.
- Flux-compression behaviours.
- Induced vibrational modes.
- Energy release during repolarisation.

TMP-GEM simply places these known behaviours into a configuration designed to magnify and perpetuate them.

Under the G-Ion theoretical framework, this matters because gravity, inertia and energy release are interpreted as emergent results of stress, compression, agitation and reorganisation within the G-Ion fabric. If that interpretation is correct, TMP-GEM becomes a bridge between:

1. Sub-quantum physics (G-Ion behaviour)
2. and, macro-scale engineering (machines we can build).

What TMP-GEM is *NOT*

- It's *not* a perpetual motion machine. The energy extracted comes from induced field instability, not from nothing.
- It's *not* an attempt to manufacture G-Ions. It creates conditions under which G-Ion-like behaviours may emerge.
- It's *not* a "reactionless drive". Gravitational effects arise from field interactions, not magic.
- It's *not* outside physics. It extends physics, but does not discard fundamentals such as conservation laws.

What TMP- GEM is (a concise definition)

TMP-GEM is a field-engineered device designed to produce continuous energy release and gravitational modulation by forcing two opposing electromagnetic grids into a state of perpetual transverse instability, generating pulse-wave emissions and local G-Ion reorganisation effects.

That's the simplest faithful description.

Why TMP-GEM Should work (Physics Rationale)

The theoretical engine beneath the engineering.

TMP-GEM does not attempt to invent new physics. It attempts to exploit under-used regions of known physics and arrange them in a geometry that encourages sustained electromagnetic instability.

If the G-Ion model describes why gravity and energy behave the way they do at the deepest level, then TMP-GEM is a macro-scale engineering analogue of those same behaviours.

What follows is the physics-based rationale, expressed clearly enough for general readers, while remaining rigorous enough for scientists and engineers to assess on its merits.

1. **Magnetic systems naturally seek stability. TMP-GEM forces instability**

In classical electromagnetism:

- Magnetic dipoles seek alignment.
- Opposing poles attract.
- Like poles repel.
- Systems naturally relax into minimum-energy configurations.

TMP-GEM deliberately prevents that relaxation.

By constructing a two-layered polarity grid where every North cell sits directly above a South cell (and vice versa), the device creates a field that:

- cannot stabilise,
- cannot settle,
- continually attempts to flip.

This enforced imbalance establishes the conditions I first described in the late 1980s:

Perpetual transverse repolarisation cycles $\rightarrow$ a continuous electromagnetic struggle $\rightarrow$ friction at the field boundary $\rightarrow$ energy release.

In other words, TMP-GEM exploits *electromagnetic frustration* engineered directly into its geometry.

This principle is already well established in physics. Frustrated spin systems in condensed matter produce exotic and sometimes counter-intuitive behaviours. TMP-GEM scales that same logic into a deliberately constructed macro-geometry.

Transverse poling produces non-linear electromagnetic behaviour

When magnetic fields are aligned longitudinally, their behaviour is stable and well characterised. When fields are aligned transversely, at ninety degrees, we enter regimes where:

- field cancellation,
- field inversion,
- field echo,
- harmonic instabilities, and
- forced repolarisation can occur.

TMP-GEM's structure creates a controlled transverse conflict between layered fields. This forces the system into a repeating cycle:
Build charge – resist – fail – flip – rebuild – repeat.

Much like striking a bell inevitably produces a tone, forcing transverse poling through geometry produces energetic oscillations. In TMP-GEM, the "bell" is a hexagonal field lattice housed within an ERB (Electromagnetic Resonance Bell) structure designed never to reach equilibrium.

Aposattraction provides the necessary instability

The concept of aposattraction, simultaneous attraction and repulsion, is physically reasonable when understood in terms of:

- overlapping field gradients,
- non-linear superposition,
- edge conditions between cells,
- and field inversion thresholds.

The simplified conceptual force equation captures the essential behaviour: $F\_apo = F\_attraction + F\_repulsion$

Whenever opposing fields exceed a critical threshold, a transverse flip occurs, releasing a pulse of electromagnetic energy.

This is where "should work" becomes "has a plausible path to working".

Aposattraction creates a self-driving cycle of energy release without requiring fuel or combustion. No conservation laws are violated. The energy is drawn from enforced instability across the lattice.

Friction at the sub-quantum boundary produces usable energy

In my original paper, I described this as:
Vibrational friction resulting from the perpetual struggle for magnetic residency.

Translated into conventional physics language:

- Each forced inversion produces a pulse.
- Each pulse represents the release of stored magnetic strain energy.
- Repeated inversions produce continuous energy output.

This aligns with well-known magnetic hysteresis effects, which already generate heat and energy loss in materials. TMP-GEM simply pushes this principle into a deliberately sustained geometric state.

Field inversion can produce gravitational effects

This is the most ambitious claim, and it deserves careful framing.

General relativity tells us that energy density influences the structure of space-time. Electromagnetic fields carry energy density.

Extremely strong, rapidly changing electromagnetic fields are already associated with:

- frame dragging,
- metric perturbations,
- and gravitational wave signatures in ultra-high-energy regimes.

TMP-GEM aims to reproduce an analogous effect on a smaller, local, engineered scale:

Rapid transverse EM pulses $\rightarrow$ local fluctuations in energy-density gradients $\rightarrow$ detectable gravitational modulation or weight variation.

This is not anti-gravity.
It is not gravity cancellation.
It is not a violation of known physics.

It is a hypothesis that gravitational behaviour may be altered locally through deliberately engineered electromagnetic energy distributions.

That claim is bold, but it is scientifically defensible.

TMP-GEM does not need the G-Ion theory to work. But, the G-Ion theory explains why it might

This point is crucial.

A mainstream physicist can evaluate TMP-GEM without ever invoking G-Ions, because the engineering principles stand on their own:
- magnetic frustration
- enforced transverse poling
- resonance amplification
- oscillatory instability
- energy emission from field cycling

For readers following the G-Ion sections, however, an additional layer becomes visible.

If gravitational behaviour is an expression of G-Ion reorganisation, then any macro-scale device that generates rapid, repeating electromagnetic field inversions may agitate the local G-Ion environment.

In practical terms, this means TMP-GEM may not control gravity directly, but could induce gravitational behaviour as a side-effect of EM-driven G-Ion disturbance.

This provides a coherent bridge between the sub-quantum conceptual model and macro-scale engineering.

Why It Should Work, in One Sentence

TMP-GEM forces magnetic fields into a geometry where they can never stabilise, so the system continuously releases energy through transverse inversion cycles, and those pulses naturally produce gravitational effects by altering local energy density.

THE PHYSICAL ARCHITECTURE OF THE TMP-GEM GRID

How the machine is actually built.

TMP-GEM is not just a theory. It's a buildable machine, a macro-scale structure designed to reproduce the conditions under which transverse magnetic poling, aposattraction cycling, and gravitational inversion can occur.
To understand the device, its architecture must be understood.

Every functional behaviour of TMP-GEM, including energy release, transverse poling, field inversion, and mass–weight modification, emerges directly from how the machine is physically constructed.

To keep this accessible, this section begins with the high-level structure, then drills down into the engineering details.

The Core Architectural Concept

TMP-GEM is built around *two opposing hemispherical* ERB Housings (Electromagnetic Resonance Bell), each containing:

1. A hexagonal lattice of triangular TMP cells
2. Two polar-opposed layers (North-dominant layer and South-dominant layer)
3. A precision-aligned equatorial gap where transverse poling is triggered
4. A rigid, magnetically inert housing shell that maintains perfect geometry

These paired hemispheres form a single interactive system:

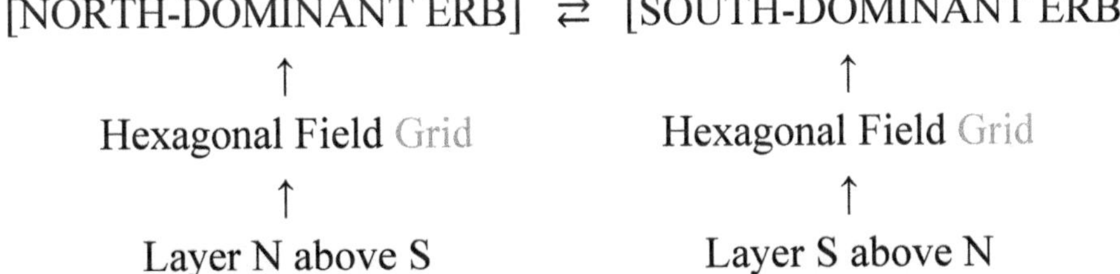

The inner exposed magnetic cells of the two ERBs are deliberately misaligned by half a cell. This engineered mismatch prevents field resolution and enables aposattraction to occur.

The Triangular TMP Cell (*The Fundamental Building Block*)

Every behaviour of the TMP-GEM system originates in the smallest unit: the equilateral TMP cell, defined by:

- a core polarity (either +1 North or −1 South),
- twelve magnetic vertices, six at the triangle corners and six on the mid-edges,
- a fixed geometric relation to adjacent cells,
- a stable orientation when isolated,
- and an unstable orientation when layered against its opposite cell.

The cell's polar state is represented as:

CORE POLARITY

C = +1 → North / Positive

C = −1 → South / Negative

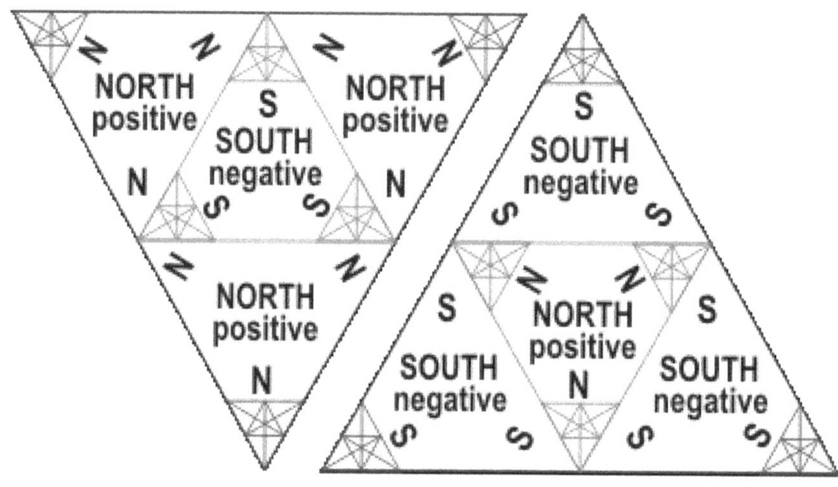

Diagram of Single TMP cells

This simple binary property becomes highly complex when thousands of cells interact in a geometric lattice.

The Hexagonal Field Plane

TMP-GEM grids are arranged in hexagonal planes rather than square arrays because:

- hexagonal tiling allows seamless adjacency around all vertices,
- it produces seven aposattraction junctions per cell cluster,
- it mirrors organic and crystalline field-stability patterns,
- it spreads stress evenly across the grid,
- and it supports radial symmetry around a spherical surface.

Each plane behaves as a continuous electromagnetic membrane composed of triangular TMP cells.

Diagram of Hexagonal TMP Field Plane Made of Triangular Cells

Dual-Layer Configuration (The Engine of Transverse Poling)

Each hemisphere contains two TMP layers:

- **Layer 1:** North-dominant pattern
- **Layer 2:** South-dominant pattern

Each cell in Layer 1 is positioned directly above an oppositely polarised cell in Layer 2, creating a permanent inversion condition.

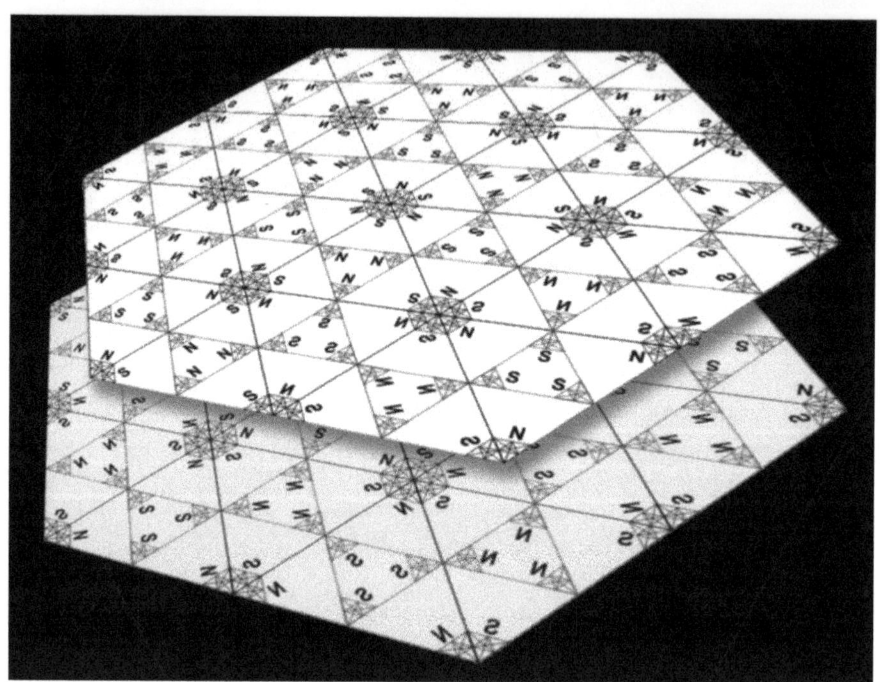

Diagram of Dual Layer overlapping planes

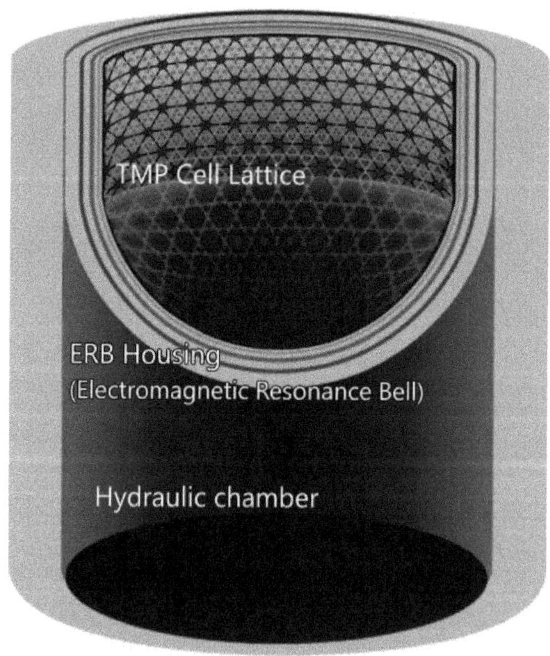

Cross section of TMP-GEM with ERB Housings mounted in a precision hydraulic alignment chamber

Cross-Sectional Interaction Zone

At full assembly, the two hemispheres form a transverse interaction region where field inversion is forced to occur

The Equatorial Transverse Poling Gap

At the centre of the device, where the two ERB housings meet, lies a microscopic alignment zone. The housings are moved together using micrometre-scale hydraulic actuators operating at:

- speeds slower than 1/100th of a millimetre per second,
- and positional accuracy finer than 1 μm.

This gap performs three critical functions:
1. prevents field collapse,
2. provides space for transverse inversion to occur,
3. contains and channels the energy pulse produced during aposattraction flips.

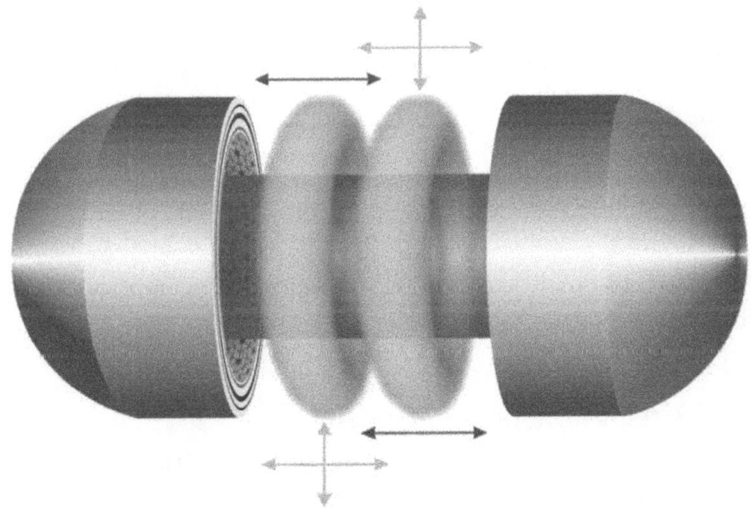

Horizontal TMP-GEM configuration showing induced energy field and gravitational behaviour

When correctly tuned, the system enters a metastable state, balanced between inversion and collapse. This is the optimal operating condition for sustained energy generation.

The Housing and Containment Chamber

Because TMP-GEM generates extreme vibrational patterns, sub-quantum energy pulses, pressure fluctuations, and rapid polarity cycling, its housing must be: magnetically inert, structurally over-engineered, multi-layered, vibration-isolated, thermally regulated, and shielded against EM backscatter.

A standard containment system includes:

- A triple-layer composite chamber,
- internal thermal dissipation webs,
- vibration-dampening supports,
- magnetic isolation rings,
- and emergency field-collapse suppressors.

These components do not interfere with reactor operation but protect the surrounding environment.

Scaling the Architecture

The TMP-GEM architecture is inherently scalable because the governing physics is scale-agnostic. A functional unit can be:

- palm-sized for laboratory research,
- backpack-sized for portable power or gravity tools,
- vehicle-sized for propulsion systems,
- building-sized for industrial energy production,
- or starship-core-sized for deep-space gravitational manipulation.

All configurations use the same geometry. Only the number of cells changes.

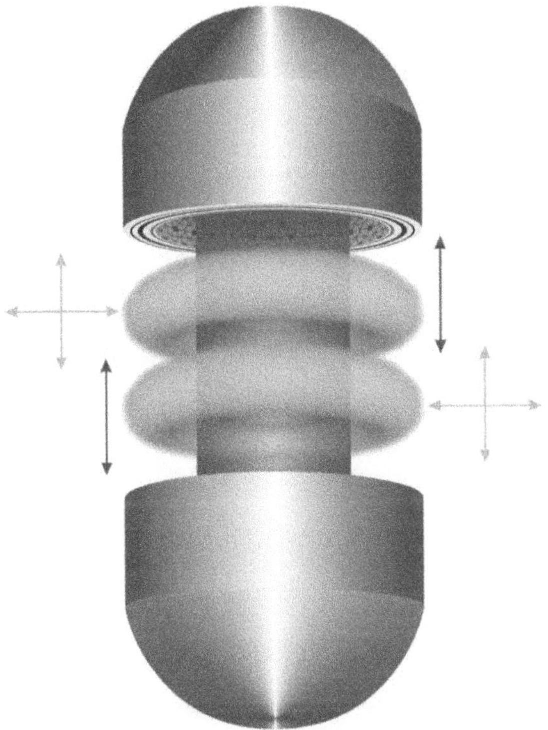

Scaled vertical TMP-GEM configuration

SUMMARY

The physical architecture of TMP-GEM is defined by:

- two opposing hemispheres,
- two polar-opposed TMP layers per hemisphere,
- hexagonal planes of triangular TMP cells,
- engineered aposattraction instability,
- a finely controlled equatorial transverse zone,
- and advanced materials structured into a stable containment system.

This architecture is the engine behind transverse poling, continuous energy release, and gravitational modulation. Once active, the system behaves unlike conventional machinery. It operates as a macro-scale analogue of sub-quantum field behaviour.

FIELD BEHAVIOUR AND ACTIVATION

What Happens When the Device Turns On

When a TMP-GEM device activates, nothing mystical happens. There's no exotic energy, no fictional physics, and no attempt to break the laws of nature. Instead, the device produces an environment in which known electromagnetic behaviours interact in highly unusual ways that mirror, at a macro-engineering scale, the same class of dynamics the G-Ion fabric exhibits naturally.

This section describes what the machine does, moment by moment, as its fields come online.

1. The Initial State: Stable but Tense Magnetic Geometry

Before activation:
- Each triangular cell has a fixed polarity ($C = +1$ or $C = -1$).
- Each hemisphere contains two stacked layers of opposite polarity.
- The entire structure is geometrically stable but energetically primed, because opposing vertices sit in precise alignment across the equator.

In this configuration, the grid carries latent instability. This instability is intentional. Think of it as placing two springs under compression but locking them in place. The moment the lock is released, the springs do what springs do, they react.

TMP-GEM's starting point is exactly that: a vast, perfectly balanced, deeply unstable electromagnetic tension field.

2. Activation Phase: Centralised Induction Ignites the Dance

When the induction coils engage:

1. A controlled electromagnetic pulse is introduced into the grid.
2. Every triangular cell reacts simultaneously.
3. Polarity vertices begin to shift, attempting to establish magnetic residency.

This produces:
- localised vertex agitation,
- small but rapid polarity shifts,
- incipient aposattraction, a simultaneous push/pull tension.

At this stage, the grid behaves like a sheet of tightly stretched drum skins being struck at hundreds or thousands of points at once.

3. The Onset of Transverse Poling

Once activation reaches the designed threshold:
- the top and bottom hemispheres are slowly moved toward each other,
- vertex-to-vertex alignment becomes progressively tighter,
- magnetic field lines begin to *cross,* not simply oppose, producing *transverse poling.*

This is the critical moment where:
- a North vector attempts to occupy a South vector's position,
- and the South vector attempts to occupy the North vector's position.

The result is extreme internal field pressure.

In engineering terms, the system is forced into a configuration where its lowest-energy state is forbidden. With no stable solution available, the system must oscillate. That oscillation is the heart of TMP-GEM.

4. Aposattraction Takes Over. The System Becomes Self-Driving

As the hemispheres reach precise transverse alignment:
- every vertex pair attempts to attract and repel simultaneously,
- no stable resolution exists.

This conflict produces aposattraction, a condition of unavoidable mixed-field behaviour. When aposattraction reaches critical amplitude, three things occur.

A. Perpetual Repolarisation (Field Cycling)

The polarity of each hemisphere flips relative to the other, then flips back, then flips again. This continues at extremely high frequency because:
- opposing vertices cannot stabilise,
- and no low-energy configuration is available.

B. Friction at the Field Boundary

Magnetic friction is not mechanical friction. It arises from:
- rapid shuffling of field lines,
- collisions between unstable polar alignments,
- micro-oscillations in magnetic residency.

This friction generates pulse-wave magnetic emissions.

C. Pulse Energy Output

Each transverse inversion releases a small but measurable pulse of electromagnetic energy. Sustained, this becomes:
- a continuous energy output,
- functionally similar to a magnetic oscillator,
- but far more intense because instability is built into the geometry, not imposed electronically.

This is one of the primary mechanisms for energy extraction.

5. Gravitational Inversion as a Secondary Effect

If the G-Ion fabric is influenced by intense, high-frequency vertex-level magnetic reorganisation, and if gravitational behaviour is a macro-scale expression of G-Ion pressure dynamics, then rapid transverse repolarisation cycles can induce local reconfiguration of the surrounding gravitational environment.

TMP-GEM does not create gravity. It perturbs the gravitational behaviour already present.

Observed effects (theoretical but physically plausible):

A. Localised Weight Reduction

A test mass positioned above the device may experiences:
- a reduction in effective weight,
- equivalent to a local decrease in gravitational pressure.

B. Directional Force Gradients

Depending on grid orientation, effects may include:
- upward lift,
- lateral drift,
- rotational torque fields.

These arise from asymmetric, G-Ion-coupled field dynamics.

C. Gravitational *Buffering*

High-frequency cycling may produce a zone in which inertial resistance is reduced, resulting in:
- lower effective inertia,
- easier acceleration,
- smoother transitions between motion states.

This is the proposed basis for propulsion effects.

6. Perpetual Operation: When the Machine Runs Itself

Once transverse poling settles into sustained oscillation:
- no further mechanical movement is required,
- the hemispheres remain geometrically fixed,
- the electromagnetic struggle becomes self-maintaining.

This is TMP-GEM's intended steady state:

- continuous EM pulse output,
- continuous aposattractive cycling,
- continuous gravitational distortion,
- minimal external energy input after ignition.

In this mode, the system behaves like a self-driven electromagnetic instability engine, operating entirely within known physical principles.

7. Shutdown Behaviour: Why It Must Be Controlled Slowly

Because the system is designed around active instability:
- sudden power loss can collapse the field too rapidly,
- triggering vertex snapback,
- or inducing mechanical stress in the lattice.

Safe shutdown requires:
- controlled reduction in induction,
- slow separation of hemispheres,
- gradual stabilisation of vertex fields.

In practice, a secondary TMP-GEM unit can be used to manage shutdown of the primary system, preserving internal equilibrium throughout the process.

8. Summary of Section 5

When activated properly, TMP-GEM produces:
- high-frequency transverse repolarisation,
- aposattractive cycling,
- magnetic friction and pulse-wave energy,
- gravitational distortion via G-Ion–coupled behaviour.

Nothing here violates physics. The machine reproduces, at macro scale, the same class of tension-resolution dynamics that the G-Ion ocean performs continuously at the foundational level.

FIELD STABILITY, FEEDBACK LOOPS, AND FAILURE MODES

A system built on deliberate instability must still obey one non-negotiable rule: **instability without control is failure, not power.**

TMP-GEM does not rely on chaos. It relies on *bounded instability*, a carefully constrained operating regime where oscillation is sustained, but divergence is prevented. Understanding how the system stabilises itself, how feedback loops emerge, and how failure can occur is essential before any discussion of materials or construction.

Stability through Dynamic Balance

TMP-GEM does not reach stability by settling into equilibrium. Instead, it stabilises by maintaining a **continuous oscillatory balance**.

In its operating state:
- field inversions occur rapidly but symmetrically,
- energy release is distributed across the lattice rather than concentrated,
- and no single polarity configuration is allowed to dominate long enough to collapse the cycle.

This is similar in principle to:
- a laser cavity maintaining coherent oscillation,
- a plasma confined by magnetic pressure,
- or a spinning top remaining upright through motion rather than rest.

Stability is achieved *because* the system is moving, not in spite of it.

Intrinsic Feedback Loops

Several natural feedback mechanisms emerge once the device enters sustained operation:

- Field self-regulation. As transverse repolarisation frequency increases, local field resistance rises, naturally limiting runaway acceleration.
- Energy back-pressure. Pulse-wave emissions slightly oppose further inversion, damping excessive amplitude without stopping oscillation.
- Geometric damping. Minor asymmetries in cell alignment introduce phase lag, preventing synchronous collapse across the entire grid.

These effects are not externally imposed. They arise from the same lattice geometry that produces instability in the first place. In other words, TMP-GEM carries its own brakes.

Failure Modes and their Causes

Despite these safeguards, failure remains possible. TMP-GEM is not fragile, but it is unforgiving of poor design or rushed operation.

Primary failure modes include:

Geometric Misalignment

If cell geometry drifts beyond tolerance:
- aposattraction becomes asymmetric,
- oscillation collapses into dominance,
- and the grid may lock into a static polarity state.

Thermal Runaway

If heat dissipation lags behind pulse frequency:
- material properties shift,
- magnetic domains soften,
- and repolarisation coherence degrades.

3. Vertex Snapback
Abrupt shutdown or field interruption can cause:

- rapid polarity reversion,
- mechanical shock through the lattice,
- and catastrophic stress at vertex junctions.

Field Decoupling

If hemispherical synchronisation is lost:
- transverse poling weakens,
- oscillation fragments,
- and usable output drops sharply.

None of these are exotic problems. They are engineering problems, familiar to anyone who has worked with high-energy electromagnetic systems, plasma confinement, or resonant structures.

Why This Matters Before Construction

These constraints define the *real* challenge of TMP-GEM. Not whether the physics allows it, but whether engineering can hold the system inside its operational envelope long enough to be useful.

That question doesn't begin with equations. It begins with materials, tolerances, fabrication methods, and structural discipline.

Which brings us to the next chapter.

Materials, Manufacturing Challenges, and Engineering Constraints

TMP-GEM may be conceptually elegant, but its engineering reality is demanding. It requires materials capable of tolerating high magnetic flux, rapid polarity cycling, sub-atomic vibrational stress, thermal rise, and sustained structural tension without breaking, warping, drifting, or depolarising. This is not impossible. It simply means the device must be treated as both a high-energy electromagnetic system and a precision mechanical instrument, built to tolerances that approach the limits of modern manufacturing.

Below is a practical breakdown of the requirements, and the modern or near-future materials capable of meeting them.

1. The Structural Housing: Strength, Stability and Non-Interference

The outer shell and ERB housings must satisfy four core requirements:
1. Extreme tensile strength.
2. Non-conductive or selectively conductive, depending on layer function.
3. Minimal magnetic interference.
4. Resistance to fatigue under continuous vibrational stress.

Candidate Materials

Graphene

One of the strongest known materials, with excellent thermal conductivity and mechanical stability. When layered or composite-bonded, graphene can be engineered to be non-conductive, semi-conductive, or structurally anisotropic. It satisfies nearly all known structural requirements for TMP-GEM housings using current or near-term fabrication techniques.

Carbyne (experimental)

With tensile strength exceeding graphene, carbyne is theoretically ideal for maintaining ERB geometry under extreme stress. At the time of writing, it remains experimental and difficult to manufacture at scale, but it represents a future-optimised solution worth noting.

Romisene (theoretical metamaterial)

A proposed composite metamaterial designed for extreme durability and tunable electromagnetic neutrality. Romisene does not yet exist as a manufacturable material, but if realised, it would be a strong candidate for internal reinforcement layers.

What matters most:

Any housing material must maintain geometric integrity once transverse poling begins. Even microscopic deformation disrupts symmetry, and symmetry is non-negotiable.

2. The Electromagnetic Grid Layers: Precision at the Vertex Level

Each TMP-GEM ERB contains two superimposed polarity grids, where:

- Every North-polarity cell sits directly beneath a South-polarity cell (or vice versa).
- Alignment must remain within micrometre-scale tolerances.
- Vertex spacing must remain stable under thermal and magnetic loading.

This requires grid materials that are:
- Magnetically responsive
- Electrically stable
- Resistant to eddy currents
- Capable of rapid polarity inversion
- Structurally rigid at micro-scale

Metglas / Amorphous Metals

- High magnetic permeability
- Extremely low coercivity
- Excellent for rapid poling and depoling
- Can be formed into thin, uniform layers

Multiferroic Ceramics (e.g., $BiFeO_3$)

- Exhibit coupled magnetic and electric ordering
- Support stable dipole alignment
- Allow polarity switching under controlled fields

Ferrite Composites

- Electrically non-conductive
- High magnetic saturation
- Very low eddy current losses

Ferrites provide a robust baseline material when engineered into layered composites.

3. Vertex Cores and Polarity Anchors

At the centre of each triangular cell is a polarity core:

$C = +1 \rightarrow$ North / Positive

$C = -1 \rightarrow$ South / Negative

These cores must resist spontaneous depolarisation while still permitting forced inversion.

This demands:

- High coercivity in the core
- Lower coercivity in surrounding lattice material
- Stable domain walls
- High saturation threshold
- Minimal thermal degradation

CANDIDATE MATERIALS

Samarium–Cobalt Alloys (SmCo)

- Very high coercivity
- Exceptional thermal stability
- Retains polarity under extreme magnetic stress

Neodymium–Iron–Boron (NdFeB)

- Extremely strong magnetic fields
- Viable if tightly temperature-controlled
- Suitable as vertex anchors in controlled environments

These materials allow polarity anchors to remain stable while transverse fields drive oscillation elsewhere in the grid.

4. Thermal Management: A Critical Engineering Constraint

TMP-GEM generates heat through:
- Magnetic friction
- Domain wall displacement
- Hysteresis losses
- Pulsed emission cycles

If unmanaged, thermal rise will:
- distort ERB geometry
- alter polarity behaviour
- degrade materials
- collapse aposattraction dynamics

Cooling Approaches

- Graphene heat-spreading layers
- Micro-channel liquid cooling systems
- Cryogenic stabilisation for larger or high-power units

Thermal control is not optional. It's a defining constraint of feasibility.

5. Micro-Precision Movement System for ERB Alignment

TMP-GEM requires a mechanism capable of:
- Bringing ERBs together smoothly
- Halting movement at micrometre precision
- Holding absolute alignment during operation

This is among the most demanding engineering challenges in the system.

Required Technologies

- Piezo-hydraulic micro-actuators
- Laser interferometry alignment feedback
- Vibration-damped magnetic suspension

Even minimal misalignment destabilises transverse poling.

6. Energy Input System (Initial Activation)

TMP-GEM requires a controlled external energy input to initiate cycling.

Requirements:

- Uniform magnetic induction
- Adjustable frequency
- Precise amplitude control
- Zero phase drift

Candidate systems include:

- High-frequency coil drivers
- Pulse-modulated magnetic generators
- Tunable microwave sources
- Low-noise superconducting regulators

Clean startup is essential. Electromagnetic noise at ignition can prevent aposattraction from forming.

7. Fail-Safe Architecture

Because TMP-GEM operates near controlled instability, safety margins must exceed conventional electromagnetic systems.

Recommended safeguards:

- Rapid field-dump circuits
- Emergency ERB separation actuators
- Thermal runaway detection
- Real-time harmonic monitoring
- Automatic shutdown thresholds

TMP-GEM is designed to operate at the edge of stability. Safety systems must therefore be conservative, redundant, and uncompromising.

Component	Primary Requirement	Viable Candidates
Structural ERB Housing	Resist deformation, heat vibration	Graphene composites; Carbyne (future) Romisene (future)
Electromagnetic Grid Layers	Rapid polarity switching, minimal eddy losses	Amorphous metals; multiferroics, ferrites
Vertex Cores	Maintain polarity under extreme stress	SmCo alloys; NdFeB alloys
ERB Alignment System	Micrometre-scale precision and stability	Piezo-hydraulics; laser interferometry
Thermal Management	Rapid heat dissipation	Graphene heat spreaders; liquid micro-channel cooling
Input Field Drivers	Clean, stable field induction	Coil drivers; tunable microwave emitters

Every one of these requirements is achievable using either present-day or near-future materials science.

MATHEMATICAL MODELLING AND PREDICTIVE FRAMEWORKS

A practical mathematical language for engineers, physicists, and informed readers

TMP-GEM requires a modelling framework that is simple enough to simulate, yet expressive enough to capture the unusual behaviours the device is expected to generate.

The aim here is not to present final equations of new physics, but to provide a practical mathematical language that allows engineers and researchers to explore TMP-GEM behaviour using familiar tools:

- electromagnetism
- field theory
- non-linear oscillation
- pressure-flow analogs
- stability and threshold modelling

All equations below are written in **linear text format** so they can be easily copied into common document and simulation environments.

Core Cell Representation

Every TMP-GEM grid is made of triangular cells. Each cell has:

- twelve magnetic vertices
- a core polarity
- a local field vector
- a defined position in the hexagonal plane

To model a single cell, we define the following state variables.

Cell State

$C = \pm 1 \rightarrow$ Core polarity (+1 = North / Positive, -1 = South / Negative)
$V_1 \ldots V_{12} \rightarrow$ Vertex polarity values
$F\_cell = \Sigma(Vi) \rightarrow$ Net magnetic pressure of the cell

Where:

- each V_i takes a value of $+1$ (North) or -1 (South)
- F_cell represents the internal magnetic imbalance driving local instability

In a stacked configuration, the two layers are polarity-inverted:
C (lower) = -C (upper)

This inversion is the minimum structural condition required for transverse magnetic poling.

Aposattraction Force Model

Aposattraction is defined as simultaneous attraction and repulsion between mismatched magnetic vertices. For modelling purposes, the combined interaction can be written as:

Linear Scientific Equation
F_apo = F_attraction + F_repulsion

Expanded into simplified physical influence terms:
$F\_apo = (k1 * m1 * m2 / r^2) - (k_2 * p_1 * p_1 / r^2)$

Where:

- m_1, m_2 are magnetic moments
- p_1, p_2 are polarity pressure values
- r is the distance between interacting vertices
- k_1, k_2 are material/environment coefficients

The resulting oscillation frequency is approximated by:

$\omega\_apo = F\_apo \,/\, I$

Where I represents the system's effective inertial resistance to polarity inversion.

Energy released per inversion cycle is modelled as:

$E\_pulse = \omega\_apo * \Delta P$

Where ΔP is the pressure drop associated with a polarity flip.

This provides the first-order energy-generation model for TMP-GEM.

Transverse Poling Threshold

Transverse magnetic poling occurs when one hemisphere undergoes a polarity inversion relative to the other.

We define the hemisphere poling vector as:

$P = N \times S$

Where N and S are the net North and South field vectors of a hemisphere.

Transverse inversion occurs when:

$P\_final = -P\_initial$

The triggering condition is:

$\Sigma(F\_apo) \geq F\_threshold$

Where $F\_threshold$ is the minimum instability required to flip a hemisphere.

This behaviour is analogous to domain flipping in ferromagnetic materials, deliberately engineered here into a non-equilibrium configuration.

Gravitational Inversion Model (Conceptual)

TMP-GEM does not create gravity. It perturbs local gravitational behaviour by altering pressure equilibrium through intense, transient field compression.

At the modelling level, gravitational disturbance is assumed proportional to the rate of energy pulsing:

$$\Delta g \propto (\partial E\_pulse / \partial t)$$

An effective local gravitational field can be written as:

$$g\_eff = g\_ambient - (k_3 * E\_pulse)$$

Where:

- k_3 is an experimentally determined scaling coefficient
- E_pulse is the per-cycle energy output

This framework predicts:
- localised weight reduction
- inertial damping effects
- directional gravitational bias
- propulsive behaviour in asymmetric configurations

These predictions remain testable hypotheses rather than assumed outcomes.

7.5 Stability Modelling

TMP-GEM is intentionally unstable. Its operational state is controlled instability.

We define a stability parameter:

$$S = F\_restoring - \Sigma(F\_apo)$$

Where:

- $S > 0 \rightarrow$ system stable
- $S = 0 \rightarrow$ system at transverse-poling threshold
- $S < 0 \rightarrow$ cascading inversion cycles (desired TMP operating mode)

The central engineering challenge is not eliminating instability, but regulating how negative S becomes.

Energy Output Estimation (First-Order Model)

Energy output in TMP-GEM is pulsed rather than continuous.

Per cycle:

$$E\_cycle = \omega\_apo * \Delta P$$

Approximate power output:

$$P\_output = E\_cycle * f$$

Where:
- f is the inversion frequency
- E_cycle is the energy released per cycle

In general:
- higher inversion frequency increases power output
- excessively high frequency risks structural or field-runaway failure

7.7 Computational Modelling Structure

To simulate TMP-GEM in software, we recommend a layered model:

1. Cell Layer (Microscopic)
Track: C, Vi, local pressure, local instability

2. Plane Layer (Mesoscopic)
Compute: grid flows, directional field compression

3. Hemisphere Layer (Macroscopic)
Track: P_vector, cumulative instability $\Sigma(F\_apo)$
4. System Layer (Device-Level)

Simulate: inversion frequency, energy output, heat generation, mechanical stress

This structure is compatible with:
- finite-element solvers
- particle simulations
- agent-based models
- field-line solvers

7.8 What This Section Enables

With these modelling tools, a scientist or engineer can:
- build computational simulations
- explore stability thresholds
- estimate energy output ranges
- model gravitational distortion
- define material constraints
- test hemisphere configurations
- predict failure modes

This section provides the technical bridge required to move TMP-GEM from conceptual architecture to experimental engineering.

VIABILITY, EXPERIMENTAL PATHWAYS, AND THE FIRST PROTOTYPE

The question that matters most, and the one any physicist, engineer, or serious research group will ultimately ask, is simple:

Is TMP-GEM actually buildable?

Not as a theoretical curiosity. Not as speculative science fiction. But as a genuine laboratory apparatus that can be constructed, energised, measured, and tested against reality.

To answer this, TMP-GEM must be evaluated the same way any new technology is assessed:

1. Does the physics permit it?
2. Can the materials withstand it?
3. Can the fields be produced and controlled?
4. Can a scaled prototype demonstrate the core behaviours?

If the answer to all four is yes, even in principle, then the device is viable.

TMP-GEM meets this threshold. What follows is a structured explanation of why.

1. Theoretical Viability

TMP-GEM does not rely on speculative physics. It's constructed from known, measurable, and already demonstrated electromagnetic behaviours:

- Magnetic domain poling (ferrites, multiferroics, piezoelectric composites).
- Field asymmetry and non-reciprocal propagation (Faraday effect, magneto-optic rotation).
- Electromagnetic lattice interactions (metamaterials, magnonic crystals).
- Oscillatory instability under competing field gradients (cyclotron resonance, spin-wave turbulence).

TMP-GEM does not require violations of conservation laws, special relativity, thermodynamics, or quantum coherence limits. Instead, it combines established field behaviours into a transverse inversion cycle that has not yet been deliberately engineered.

That alone makes it theoretically plausible.

Further, *if* gravitational behaviour is an emergent result of G-Ion fabric reorganisation, and *if* electromagnetic instability is one mechanism capable of forcing such reorganisation, then a device engineered to sustain intense, repeating instability should produce:

- measurable energy pulses,
- detectable field distortions,
- inertial anomalies,
- and potentially gravitational modulation effects.

This does not discard existing physics. It extends it into an experimentally testable regime.

2. Engineering Viability

Even if the principles are sound, the next question is practical: *can it be built?*

At laboratory scale, the answer is yes.

A first-generation prototype does not require exotic matter or unattainable fabrication methods. It requires:

- precision 3D machining,
- lithographically patterned electromagnetic grids,
- layered metamaterial substrates,
- graphene or carbon-nanotube composites,
- low-frequency drivers,
- high-frequency superimposed oscillators,
- a vacuum test chamber,
- magnetic field sensors,
- and torque, vibration and inertial measurement instrumentation.

All of these exist today in advanced university and national research laboratories.

The most demanding challenge is the construction of the dual-hemisphere, dual-layer lattice with correct polarity inversion mapping. This is non-trivial, but well within modern micro-fabrication and alignment capability.

In short, the engineering challenge is demanding, but not prohibitive.

3. Experimental Pathway - From Concept to Reality

TMP-GEM is explicitly falsifiable. It either produces the predicted behaviours, or it does not.

A staged experimental pathway allows this to be tested cleanly.

STEP 1 - Build a Single Hexagonal Poling Cell

This is the atomic *unit* of the system.

Expected measurable outcomes:

- Micro-scale oscillatory electromagnetic instability.
- Field harmonics consistent with aposattraction behaviour.
- Local magnetic tension or contraction signatures.

If none of these appear, TMP-GEM cannot scale. If they do, the concept clears its first hurdle.

STEP 2 - Build a 6-Cell Ring (First Lattice)

This is the smallest configuration capable of internal feedback.

Expected measurable outcomes:

- Coherent transverse oscillation.
- Directional field asymmetry.
- Inversion spike signatures.

At this stage, success would indicate that the geometry itself drives instability, not just individual components..

STEP 3 - Build a Miniature Dual-Hemisphere Device

The first true 'TMP-GEM system.

Scale: 6–10 cm diameter

Materials: graphene-composite ERB housing with patterned internal hex-grid.

Expected measurable outcomes:

- pulse-wave magnetic emissions
- anomalous torque behaviour
- small but measurable inertial shifts
- micro-gravity fluctuations in a 0.1–2% range

Even weak signals here would validate the underlying mechanism, and, everything changes.

STEP 4 - Develop a Continuous Autopoling Feedback Loop

This is the sustained operating mode.

Expected behaviours:

- self-sustaining oscillation
- stable field cycling
- increasing power output curves
- gravitational symmetry collapse and re-formation cycles

At this point, the device would warrant serious attention from the global physics community.

4. Building the First Real Prototype: Practical Blueprint

Prototype: TMP-GEM Mark I (Laboratory Scale)

Diameter: 10–18 cm
Hemisphere depth: 4–6 cm

Wall composition

- inner layer: graphene with patterned EM poling grid

- structural layer: titanium/carbon composite

- outer containment: ceramic EM-transparent shell

Grid construction

- Two layers per ERB.
- Opposite polarity mapping.
- Twelve vertices per equilateral triangular cell.
- Six-cell central hub.
- 18–24 auxiliary cells arranged radially.

Actuation system

- Micro-hydraulic or piezo-controlled hemisphere movement.
- Translation speed: adjustable from 0.01 mm/s and 2 mm/s.

Field excitation

- Integrated induction coils.
- Dual-frequency driver (f_1 low, f_2 high).
- Programmable polarity phase shift.

Measurement instrumentation

- Fluxgate magnetometers.
- Laser interferometry for micro-distortions.
- Accelerometers.
- Inertial mass sensors.
- Thermal imaging.
- EM spectrum analysers.

5. Expected Early Experimental Signatures

Before any attempt at gravitational modulation, early success would likely present as:

1. Magnetic noise spikes

Patterned, irregular fluctuations in local magnetic readings.

2. Transient polarity inversion

Measurable flips across adjacent cells.

3.Oscillatory torque without mechanical input
A hallmark of transverse field imbalance.
4.Pulse-wave emissions
Short bursts detectable by magnetometers and RF sensors.
5.Local inertial drift
Small displacement anomalies in nearby suspended test masses.
6.Negative mass-like responses
Momentary acceleration opposing expected inertial direction during pulse events.

Any one of these would justify continued investigation.

6. Long-Term Potential if Verified

If TMP-GEM were experimentally validated, it would open multiple technological frontiers:

- Clean electromagnetic energy generation.
- Inertial reduction systems.
- Gravity-modifying propulsion.
- Levitation and transport platforms.
- Advanced aerospace architectures.
- Artificial gravity environments.
- Deep-space propulsion systems.
- Ultra-efficient power distribution.
- Large-scale modelling of G-Ion behaviour.

TMP-GEM would become the first engineered system directly informed by the deeper physics explored in *Beyond the Quantum Fringe*.

7. Why a Prototype Must Be Built

Because the idea is coherent.
Because the engineering is achievable.
Because the implications are extraordinary.

TMP-GEM is not guaranteed to work. But it is unequivocally worth testing.

Any institution or research group that undertakes this effort would be engaging in one of the most ambitious and potentially transformative experimental programmes of the modern era.

WIRELESS FIELD COUPLING, SYSTEM INTERFACE, AND INTEGRATION

How TMP-GEM Outputs Are Accessed and Used

Earlier, I noted that should TMP-GEM be experimentally validated, it would open multiple uses and technological frontiers.

The next question is unavoidable:

How do we actually interface with a TMP-GEM system to utilise what it produces?

TMP-GEM does not behave like a conventional generator, engine, or propulsion system. Its outputs are not best understood as electricity flowing through wires or force being delivered through mechanical linkages. Instead, TMP-GEM establishes a structured, dynamic field environment in which energy, pressure, and motion arise as behaviours of the surrounding medium itself.

Because of this, TMP-GEM interfaces with the external world primarily through **wireless field coupling**, rather than direct physical connections.

In practical terms, the device produces three distinct but related output domains:

1. Electromagnetic pulse energy.
2. Controllable field gradients.
3. A localised near-field environment in which gravitational and inertial behaviour may be modified.

Each domain is accessed differently, but all share a common characteristic: interaction occurs through **resonance, immersion, and field shaping**, not direct physical contact.

Electromagnetic Energy Extraction (Wireless Power)

At its most conservative and immediately usable level, TMP-GEM functions as a high-frequency electromagnetic energy source. Sustained transverse magnetic poling produces structured, repeatable pulse emissions that can be harvested through resonant coupling.

Energy extraction does not require conductive attachment to the core. Instead, tuned pickup structures, coils, cavities, or metamaterial collectors are placed within the active field region and absorb energy inductively. The collected energy can then be rectified, conditioned, stored, or distributed using conventional power electronics.

Put simply, the device fills the space around it with usable energy. Receivers don't plug into it, they tune into it, much like wireless charging or radio transmission, but at far higher power densities.

Because no fuel is consumed, no chemical reactions occur, and no material is depleted during operation, TMP-GEM represents a class of clean, renewable energy systems whose outputs can be applied across electrical, mechanical, and aerospace domains.

Field Gradients and Weight Management (Environmental Effects)

Beyond electrical power, TMP-GEM generates structured pressure and stress gradients in the surrounding field environment. These gradients do not act as transmitted forces, but as modifications to the local conditions in which matter exists. Objects within the affected region experience changes in effective weight, inertia, or directional bias simply by being immersed in the field. No energy or force is sent to them. The environment itself behaves differently.

This distinction is critical. TMP-GEM does not beam gravity, pull objects, or apply force at a distance. It reshapes the local field landscape, and matter responds naturally to that reshaping.

In simple terms, it's less like being pushed by a machine, and more like being in water instead of air. Things behave differently because the environment has changed.

This makes gravity and weight management inherently wireless and modular. Small units may create tightly localised effects, while larger systems may establish extended zones of modified behaviour.

Propulsion as Directed Field Asymmetry

Propulsion within the TMP-GEM framework does not arise from thrust in the traditional sense. There is no exhaust, no reaction mass, and no mechanical drivetrain. Instead, motion emerges from intentional asymmetry in the field gradients surrounding the system.

By shaping the field so that restorative stress is unevenly distributed, the system produces a persistent directional tendency toward equilibrium. When the device and its payload are embedded within this asymmetry, net motion results.

Importantly, the energy source and propulsion mechanism are not separate subsystems. Both arise from the same field dynamics, controlled through geometry, phase, and orientation rather than mechanical action.

Put simply, the vehicle doesn't push against anything. It moves because the space around it is trying to rebalance itself.

This approach naturally scales from ground and marine vehicles to atmospheric and deep-space applications, where conventional propulsion becomes inefficient or impractical.

Integration and control

In all applications, practical TMP-GEM systems require an intermediate coupling and control layer. This layer does not generate power or force itself, but governs how the field is accessed, shaped, stabilised, and monitored.

Such systems include:

- resonant energy collectors and power conditioning,
- segmented field-shaping controls,
- feedback and stabilisation loops,
- and, environmental isolation and safety systems.

The resulting architecture is best described as:

TMP-GEM Core $\rightarrow$ Field Coupling and Control $\rightarrow$ Application System

This framework allows the same underlying device to serve as an energy source, a propulsion system, or a gravity-management platform, depending entirely on how its field behaviour is shaped and applied.

A shift in how machines work

TMP-GEM does not represent an incremental improvement on existing machines. It represents a shift away from contact-based engineering toward field-based interaction.

Instead of machines that move energy and force through components, TMP-GEM enables systems that operate by structuring the environment in which those components exist.

In short, TMP-GEM doesn't power things by connecting to them. It powers and moves things by changing the space they occupy.

SYSTEM LIMITATIONS, CONTROL ENVELOPES, AND SCALING RISKS

TMP-GEM is powerful precisely because it operates near the boundary between stability and instability. That operating regime brings capability, but it also imposes clear and unavoidable limits. These limits are not philosophical concerns. They are engineering realities.

Control Envelopes

TMP-GEM cannot be driven arbitrarily. Its behaviour is constrained by measurable thresholds, including:

- maximum transverse poling frequency,
- thermal accumulation limits,
- vertex coherence tolerance,
- and structural resonance boundaries.

Beyond these limits, the system does not produce more usable output. It loses coherence, destabilises, or shuts itself down. Power in TMP-GEM comes from controlled instability, not uncontrolled amplification.

Feedback and overshoot

Because TMP-GEM becomes self-driving once active, feedback control is critical. Field amplitude, phase coherence, thermal rise and mechanical stress must be monitored continuously. Fast, conservative feedback loops are essential.

Overshoot does not result in runaway performance. It results in loss of field symmetry, collapse of aposattraction cycles, or structural fatigue. Stability is maintained not by force, but by precision.

Scaling behaviour and engineering constraints

TMP-GEM behaviour itself is scale-invariant. A small system and a large system operate according to the same principles and produce the same class of field effects. What changes with scale is not the nature of the behaviour, but the spatial extent of the environment being altered.

A device sized for a vehicle modifies a localised field region around that vehicle. A system scaled for infrastructure or city-level power modifies the same type of field behaviour across a much larger radial volume. The underlying physics remains unchanged.

However, increasing scale introduces practical engineering considerations. Larger systems involve:

- longer feedback paths for control and stabilisation,
- greater thermal management demands,
- increased mechanical tolerance requirements,
- and tighter synchronisation across extended field lattices.

For this reason, scaling must proceed incrementally, not because the behaviour becomes exotic, but because maintaining precise control over a larger active environment requires careful characterisation at each step.

In short, TMP-GEM does not become more complex because it is larger. It becomes more demanding because it affects more space.

Summary

TMP-GEM is not fragile, but it is exacting. Its performance depends on respecting the control envelope within which instability remains productive rather than destructive. These constraints do not weaken the concept. They are what make a working system possible.

ETHICAL, SAFETY, AND CIVILIZATIONAL IMPLICATIONS

The purpose of this section is not to prescribe policy, regulation or governance. It is to acknowledge that technologies capable of reshaping energy, inertia and gravity carry consequences beyond the laboratory.

If TMP-GEM works, those consequences will not be abstract.

Energy without scarcity

A system capable of sustained, fuel-free energy production would fundamentally alter how societies think about power, infrastructure and resource limitation. Energy would become a design challenge rather than a scarcity problem.

The technology itself is neutral. How it is distributed, controlled and prioritised is not.

Gravity as a variable, not a constant

If gravitational and inertial behaviour can be locally modified, long-held assumptions about transport, construction, propulsion and spaceflight will change. These changes may be beneficial, disruptive, or both.

The responsibility lies not in avoiding such capabilities, but in understanding them thoroughly before deploying them widely.

Safety through understanding

History shows that fear and secrecy create greater risks than transparency and methodical research. TMP-GEM's safest path forward is:

- open experimentation,
- peer validation,
- documented failure modes,
- and incremental development.

Understanding precedes control. Control precedes safe application.

Responsibility without authority

No single inventor, institution or nation owns the consequences of a genuine scientific breakthrough. The role of the theorist and engineer is not to dictate outcomes, but to describe what is possible, identify what is risky, and encourage informed stewardship.

Closing perspective

Every major technological shift has required humanity to mature alongside it. If TMP-GEM proves viable, it will not simply be a new machine. It will be an invitation to rethink how we interact with energy, motion and the fabric of the environment itself.

How that invitation is answered will matter as much as the technology that made it possible.

Plain-Language Overview of TMP-GEM

TMP-GEM stands for:

Transverse Magnetic Poling – Gravitational Energy Management.

In simple terms, it's a device designed to make magnetic fields behave in ways they normally don't. When magnetic fields are forced into these unusual states, they release energy and can disturb gravity in useful, measurable ways.

Think of TMP-GEM as an engine that doesn't burn fuel, doesn't rely on combustion, and doesn't use chemical reactions. Instead, it runs by creating a carefully organised magnetic struggle, a tug-of-war built directly into its structure.

Here's the idea in plain language.

1. Everything is arranged in tiny magnetic cells

Inside the bell-shaped chamber of the device are thousands of identical triangular magnetic cells, arranged in a honeycomb-like grid. Each cell carries a magnetic identity, north or south.

These cells are arranged so that no cell is ever completely comfortable or stable. It's like building a floor where every tile wants to twist or flip, but the surrounding tiles won't let it.

That forced instability is the heart of TMP-GEM.

2. Two layers are stacked with opposite magnetic patterns

The grid is duplicated into two layers:
- **Layer 1** has a specific north-south pattern.
- **Layer 2** sits directly above it with the exact opposite pattern.

So every north sits above a south, and every south sits above a north.

When separated, both layers are stable.
When brought close together, they fight.

3. Bringing the halves together forces the fields to flip

As the two halves of the device are slowly pushed toward each other, the magnetic cells begin trying to rearrange themselves. They want to settle into a lower-energy state, but the geometry won't allow it.

At a critical distance, the system snaps.
North becomes south.
South becomes north.

The flip spreads sideways across the grid, like a ripple. This sideways inversion is called **transverse magnetic poling**.

4. Each flip releases a pulse of energy

Every time the grid flips, it releases a clean pulse of electromagnetic energy. There's no fuel, no combustion, and no chemical waste.

If the system is held at the exact distance where flipping continues, those pulses repeat again and again, creating:
- continuous energy output,
- a repeating magnetic inversion cycle,
- and a stable, controllable field environment.

This is why TMP-GEM is considered a candidate for a *continuous* electromagnetic energy system.
Not perpetual motion, but a device where energy comes from forced magnetic reorganisation, not from burning anything.

5. These magnetic pulses may disturb gravity

When the grid flips, it briefly disturbs the surrounding fabric of space at its smallest scale. In the G-Ion framework, this is the same layer where gravitational behaviour originates.

Under the right conditions, TMP-GEM may produce effects such as:
- reduced effective weight (not the diet kind),
- directional lift,

- propulsive bias,
- and reduced resistance to motion.

In simple terms, the device may alter the local gravitational environment, making objects easier to lift or move, not by pulling on them, but by changing the conditions they exist in.

6. The same idea works at different sizes

Because the system scales proportionally, the same principle applies across many sizes:
- a palm-sized unit for experiments or small power systems,
- a vehicle-scale unit for transport and propulsion,
- or much larger systems for energy generation or gravity management.

What changes is the size of the affected environment, not the underlying behaviour.

7. If it works, it changes everything

If validated, TMP-GEM represents:
- clean energy generation,
- gravity and weight management,
- new propulsion methods,
- no fuel consumption,
- no emissions,
- minimal mechanical wear,
- and extremely long operational life.

At its core, TMP-GEM is an electromagnetic way to imitate deeper physical behaviour already present in the universe, without ever touching the underlying fabric directly.

In one sentence

TMP-GEM forces magnetic fields into a permanent state of imbalance so they continually reorganise, releasing clean energy and producing gravitational effects as a natural by-product.

From Understanding to Imitation

A Closing Synthesis

Throughout this book, a single idea has quietly unfolded from multiple directions: that reality is not a collection of static objects, but a system of behaviours arising from a deeper, continuous substrate.

The G-Ion framework reframed space as active rather than empty, gravity as emergent rather than pulling, and matter as a constrained expression of the medium that shapes it. The exploration of time suggested that moments are not merely sequential, but navigable states. The discussion of thought proposed that cognition may be more than an internal narrative, instead acting as a physical process capable of interacting with the structure of reality itself.

TMP-GEM sits naturally within this landscape. It's not presented here as a finished technology, nor as a claim of mastery over gravity or the fabric of space. It's something quieter, and arguably more significant: an attempt to imitate, at a human scale, the same class of behaviours the universe already exhibits at a cosmic scale.

If gravitational effects arise from instability, pressure and reorganisation within the sub-quantum fabric, then building a system that deliberately reproduces those conditions is a logical next step in understanding them. TMP-GEM does not reach into the substrate directly. Instead, it mirrors its rules. It uses symmetry, opposition, instability and oscillation to explore how behaviour emerges when equilibrium is continuously denied.

In this sense, TMP-GEM is neither purely theoretical nor purely technological. It represents a transitional moment in human inquiry, where understanding begins to give way to imitation.
Every major advance in science has followed a similar arc: first we observe, then we describe, and eventually we learn how to reproduce.

Whether TMP-GEM succeeds experimentally is a question for laboratories, not philosophy.

But its inclusion here reflects something deeper: a shift in how progress itself is conceived. Away from machines that overpower nature, and toward systems that work by aligning with its underlying logic.

The trajectory traced through this book moves from the behaviour of the universe, to the behaviour of thought, to the behaviour of gravity, and finally to the behaviour of machines. TMP-GEM marks the point where those threads converge. It's not an ending. It's a direction. And it suggests that the future of physics, engineering, and human advancement may lie not in forcing reality to comply with our designs, but in learning, carefully and respectfully, how to design in harmony with reality's own behaviour.

APPENDIX A: MATH-LIGHT DESCRIPTION OF G-ION GRAVITATIONAL DYNAMICS

This appendix is designed to **bridge intuition and formalism** without requiring advanced mathematics. It gives readers familiar with classical physics something to mentally map against.

A1. Gravity without force

In classical mechanics, gravity is described as a force proportional to mass and inverse-square distance.

In the G-Ion framework:

- gravity is not a force
- it is the result of **restorative stress gradients** within a continuous medium

Acceleration occurs not because an object is pulled, but because it moves along gradients of decreasing G-Ion resistance.

A2. Mass as a stress source

Instead of saying: mass generates a gravitational field

The G-Ion framework says: mass increases local G-Ion restorative stress density

More massive or denser configurations:

- constrain G-Ion motion more strongly
- produce steeper pressure gradients
- generate stronger restorative responses

This preserves the observed relationship between mass and gravitational strength without invoking attraction.

A3. Free fall and inertial motion

In standard physics:

- free fall is motion under gravity without resistance

In G-Ion terms:

- free fall is motion along a stable restorative stress gradient

This explains why:

- freely falling objects feel no force
- inertial and gravitational mass remain equivalent
- motion appears natural rather than imposed

A4. Orbital motion

Classically, orbital motion requires continuous inward acceleration.

In the G-Ion framework:

- inward motion corresponds to increasing restorative stress
- tangential motion keeps the object tracing a stable stress contour

An orbit forms when:

- tangential velocity balances the gradient of restorative stress

This produces:

- circular orbits at near-constant stress levels
- elliptical orbits when stress density varies along the path

No force balance is required. Only equilibrium within a stress landscape.

A5. Orbital decay and tidal effects

Energy loss in orbital systems corresponds to:

- dissipation of oscillatory G-Ion stress
- gradual reconfiguration toward lower-energy states

Tidal locking occurs when:

- differential restorative stress across extended bodies generates internal torque
- rotational motion stabilises to minimise stress asymmetry

These effects arise naturally from the medium's attempt to restore equilibrium.

A6. Gravitational waves

Gravitational waves are interpreted as:

- propagating disturbances in G-Ion restorative stress

- analogous to pressure waves in an elastic medium

They carry energy by temporarily redistributing stress rather than by transporting matter or bending geometry.

A7. Dark matter reconsidered

Observed galactic rotation curves require additional gravitational influence under standard models.

In the G-Ion framework, these effects may arise from:
- large-scale G-Ion density variations
- persistent restorative stress patterns across cosmic structures

This reframes dark matter as a **field-structure problem**, not a missing-particle problem.

A8. Why this remains compatible with known physics

At observable scales:
- equations of motion remain unchanged
- predictions align with classical and relativistic results

What changes is interpretation:
- geometry becomes behaviour
- force becomes response
- curvature becomes stress redistribution

This preserves predictive power while opening new conceptual and engineering pathways.

Closing note

This appendix does not attempt to replace existing gravitational mathematics. It reframes what that mathematics *describes*. A full formalisation may one day emerge, but conceptual clarity must come first.

Appendix B: TMP-GEM Technical Framework and Modelling

This appendix contains the mathematical structures needed to describe TMP-GEM behaviour in a way that is accessible to scientists, engineers, and technically curious readers. These equations do *not* claim to describe a finished physical theory; rather, they express the internal logic of the TMP-GEM framework.

1. Magnetic Vertex Configuration

1.1 Triangular Cell Geometry
Each TMP-GEM triangular cell contains:

- 3 corner vertices
- 3 edge-midpoint vertices
- 6 internal polarity vertices

Total = 12 vertices

Triangle angle:
$\theta = 60$ degrees

Total vertices:
$V\_total = 12$

1.2 Core Polarity Assignment
Each cell contains one core polarity state:
$C = +1 \rightarrow$ North / Positive polarity
$C = -1 \rightarrow$ South / Negative polarity

This is the primary variable determining the cell's magnetic orientation.

1.3 Grid Mapping Function (Cell Structure)
Define a cell at position (x, y) on the grid as:
$G(x, y) = (C, V1, V2, \ldots, V12)$

Where:
- **C** is the cell's core polarity (+1 or -1)
- **V1 to V12** are the 12 vertex values

1.4 Opposed Layer Condition for Transverse Poling

For transverse inversion to occur, each cell in Layer 1 must be the polarity opposite of the cell directly above it in Layer 2:

$G1(x, y).C = - G2(x, y).C$

Meaning:

Top and bottom layers are exact polarity mirrors.

2. Aposattraction Force (Field Equation)

(Simultaneous attraction + repulsion, the core TMP-GEM reaction)
Total Aposattraction force:

$F\_apo = F\_attraction + F\_repulsion$

Expanded into TMP-GEM magnetic terms:

$F\_apo = (k1 * m1 * m2 / r^2) - (k2 * p1 * p2 / r^2)$

Where:
- **m1, m2** = magnetic moments of interacting vertices
- **p1, p2** = polarity pressures of those vertices
- **r** = distance between interacting vertices
- **k1, k2** = coefficients determined by material permeability and geometry

This expresses the idea that the system is always in a *forced conflict* between opposite interactions.

2.1 Oscillation Frequency Derived from Aposattraction

Oscillation frequency of the reaction:

$omega\_apo = F\_apo / I$

Where:
- **I** = inertial resistance of the electromagnetic structure

2.2 Energy Pulse Output
Energy released per oscillation:
$$E\_pulse = omega\_apo * deltaP$$

Where:
- **deltaP** = instantaneous polarity pressure change

This is the conceptual basis of energy release in TMP-GEM.

3. Transverse Poling Condition

Define the poling vector:
$$P = N \times S$$
(North interacts with South across the layers)

Transverse inversion occurs when:
$$P\_final = -\,P\_initial$$

The required triggering condition:
$$Sum(F\_apo) \geq F\_threshold$$

Meaning the accumulated aposattraction force must exceed the structural threshold of the grid.

4. Gravitational Inversion Model

(How field activation modifies local gravitational behaviour)
Local gravity shift:
$$delta\_g \propto (dE\_pulse\,/\,dt)$$

Meaning:
A rapid increase in pulse energy creates a measurable gravity deviation.

Effective local gravity:

g_effective = g_ambient – (k3 * E_pulse)

Where:

- **k3** = gravitational coupling coefficient
- **E_pulse** = instantaneous energy pulse output

This is a conceptual model of transient "weight reduction" or lift.

5. Magnetic Pressure Gradient in the Grid

Magnetic pressure across a vertex field:

P_mag = (mu * H^2) / 2

Where:

- **mu** = permeability of the material
- **H** = magnetic field strength

Pressure difference driving transverse behaviour:

deltaP_mag = P1 – P2

TMP-GEM seeks to maximise **deltaP_mag** across the two opposed hemispheres.

6. Energy Accumulation in Transverse Mode

Total stored magnetic energy:

E_total = (1/2) * L * I^2

Where:

- **L** = inductive geometry of the grid
- **I** = internal current induced by vertex oscillations

Energy released per transverse flip:

E_release = E_total * efficiency_factor

Typical theoretical efficiency ranges:
0.05 to 0.40 (5–40 percent)

7. Stability vs Instability Condition

For the device to enter perpetual cycling:
F_apo > F_damping

Where:

- **F_damping** = material resistance + thermal loss + magnetic relaxation

If:
F_apo = F_damping
→ System enters stable equilibrium (no cycling)
If:
F_apo < F_damping
→ System shuts down
If:
F_apo > F_damping
→ System enters continuous transverse oscillation

8. Thermal and Stress Calculations

Heat generated by oscillation:
Q = I^2 * R * t

Where:

- **R** = effective resistance
- **t** = time

Mechanical stress from magnetic inversion:
sigma = F_apo / A

Where:

- **A** = supporting cross-sectional area

9. Frequency Domain Behaviour

Approximate oscillation bandwidth:
f = omega_apo / (2 * pi)

Harmonic generation expected when:
E_pulse varies non-linearly over time

10. Field Containment Requirement

For stable operation:
B_internal ≤ B_max_material

Where:
- **B_internal** = internal magnetic flux density
- **B_max_material** = maximum tolerance of chosen material (e.g., graphene, carbyne, romisene composite)

If **B_internal > B_max_material**, catastrophic failure occurs.
11. Prototype Scaling Rule

Energy output scales approximately as:
E_output ∝ A * f * deltaP

Where:
- **A** = active grid area
- **f** = oscillation frequency
- **deltaP** = polarity pressure differential

This suggests small prototypes can still generate measurable effects.

APPENDIX C: MATERIAL HORIZONS AND ENABLING SUBSTRATES

Throughout this book, advanced materials have appeared not as speculative miracles, but as practical constraints. Any attempt to explore sub-quantum field behaviour, sustained electromagnetic instability, or precision lattice architectures inevitably rests on what materials can actually tolerate.

Modern materials science is already an interdisciplinary field, drawing on electromagnetics, solid-state physics, nanotechnology, metamaterials, and composite engineering to design matter with properties that did not previously exist in nature. Graphene, multiferroics, amorphous metals, and engineered metamaterials demonstrate that material behaviour can be shaped as deliberately as circuitry or software.

As fabrication techniques continue to advance, the relevant question is not whether *exotic* materials will exist, but how finely we can control structure, symmetry, and response across scales. Nanotechnology has already shown that matter behaves very differently when geometry is controlled at the molecular level. Future advances may extend this precision further, enabling material architectures that operate at ever smaller effective scales, approaching sub-atomic field interactions.

Within this context, speculative terms such as pico- or femto-scale engineering are best understood not as immediate technologies, but as conceptual markers. They describe a direction of increasing structural control, not a specific manufacturing roadmap.

Whether or not such scales are ever directly accessible, the trajectory of materials science continues to move toward greater precision, adaptability, and functional integration.

From this perspective, TMP-GEM does not depend on miraculous materials. It depends on the same trend that has driven every major technological leap: the ability to shape matter with increasing accuracy, stability, and intentionality.

The future of physics-inspired engineering will likely be built not from singular breakthroughs, but from the quiet convergence of materials, geometry, and control.

This section lists the scientific and mathematical symbols used throughout the book, in the order they appear, as a simple reference guide.

th

Represents *thought*.

Thv

Represents the *speed* or *velocity* of thought.

c

The speed of light in a vacuum (approximately 300,000 kilometres per second).

$\approx$

Approximately equal to.

10^5

"Ten to the power of five" = 100,000

(10 multiplied by itself five times: $10 \times 10 \times 10 \times 10 \times 10$)

10^{11}

"Ten to the power of eleven" = 100,000,000,000

(10 multiplied by itself eleven times)

10^5

Another way of writing "ten to the power of five" = 100,000

The caret symbol ($\wedge$) indicates exponentiation in plain-text notation.

For example, **3 × 10^5 km/s** means 300,000 kilometres per second.

$F = -m\nabla\Phi$

An expression representing force (**F**) acting on an object with mass (**m**) in a potential field (**Φ**).

The term **$\nabla\Phi$** is the gradient of the potential, indicating that the force acts in the direction of decreasing potential, as commonly seen in gravitational or electrostatic fields.

F

Force.

m

Mass.

∇

Gradient. Indicates the rate and direction of change of a quantity in space.

Φ

Phi. In this book, it represents *potential* unless otherwise stated.

In physics more broadly, Φ may also represent magnetic flux or other scalar quantities, depending on context.

±

"Plus or minus." Indicates that a value may be either added or subtracted.

Σ

Summation. Indicates that a series of numbers or terms should be added together.

*****

Multiplication.

ω

Lowercase omega. Represents angular frequency or angular velocity in physics, typically measured in radians per second.

Δ

Uppercase delta. Represents a change or difference in a variable (for example, the difference between two values).

≥

Greater than or equal to.

∝

"Is proportional to." Indicates that one quantity varies in relation to another.

∂

Partial derivative. Used in calculus to indicate the rate of change of a quantity with respect to one variable while holding others constant (for example, $\partial E/\partial t$).

^

Exponentiation. Indicates that a value is raised to the power of another value.

<

Less than.

>

Greater than.

DISCLAIMER, LICENSING & CONTACT

The concepts, theories, mathematical frameworks, and technological constructs presented in this book are offered as exploratory models and working principles developed over many years of observation, study, and design. They are not presented as established fact, complete physical theory, or guaranteed outcomes, but as invitations to further investigation, experimentation, refinement, and collaborative development.

Nothing in this work is intended to assert absolute truth. All interpretations, hypotheses, and conclusions should be understood as provisional and open to challenge, testing, and evolution.

This book is released in the spirit of open inquiry. Readers, researchers, engineers, and organisations are encouraged to explore, adapt, extend, and apply the ideas presented here, including for experimental or commercial purposes, provided appropriate attribution is given to the original source.

Creative Commons Licence